MAKE

WORLDS

Giovanni P. Gregori
and Bruno G. Gregori

The Universe out of the Box Worlds and Anti-worlds

New Foundations of Physics

Rhythms, Golden Ratio
Origin of Life, Antifragility

europe books

www.europebooks.co.uk | info@europebooks.co.uk
ISBN 9791256971787
First edition: October 2025

The Universe out of the Box
Worlds and Anti-worlds

Dedication

"Canta in un attimo il Suo eterno silenzio
Laddove non scorre nemmeno il Mio tempo:
In cima al Mio mondo, sul Colle del Vento"

"He sings in an instant His eternal silence
Where not even My time passes:
On top of My world, on the Hill of the Wind"

The explanation is as follows.

"He sings in an instant His eternal silence",
i.e., the Absolute, or God, in His timeless β-world, sings in an instant His eternal silence.
"Where not even My time passes:"
i.e., even My time does not flow in the β-world:
"On top of My world, on the Hill of the Wind",
i.e., while I live in My α-world, on the Hill where Wind flows, i.e., where every accidental and transient, non-absolute, phenomenon happens.

This poem was composed, on August 11th 2019, by a teenager (16 – 17) poet, Francesco Spadoni, to whom the concept had been very briefly explained by his grandfather G.P. Gregori. Francesco unexpectedly passed away on January 8th 2024 at the age of 21, leaving his parents, brother, grandparents, and the whole family heartbroken.

This poem expresses, in an admirable way, in three verses, the essence of the guiding concept of this volume, linked to the explanation of the existence of antimatter.

Contrary to what is generally believed, "time" is not "absolute". The "absolute" world in nature is the one without "time" and we will call it the β world. From the world β, occasionally, two mirror worlds arise: the world α where matter and "time" exist - and where we live - and the world γ where antimatter and antitime (which flows from the future towards the past) exist. Matter and antimatter, or the α world and the γ world, sometimes annihilate each other by returning to the β world.

It is thus possible to reasonably hypothesize that the deceased leave their worn out and now unusable body in the world α, and enter the timeless world β of the Absolute.

With immense love and heartbreaking regret, this volume is dedicated to the memory of Francesco Spadoni.

This same adolescent poet, when he was not yet *14* years old - in one of his compositions which he spontaneously composed and occasionally sent to his grandparents - wrote (on 15 April 2016) this sentence, which we report here with profound emotion - and which is well suited to the breadth of the problems addressed here.

"Humanity - Humanity is a great machine, perfect in its imperfection. The fuel that drives it is intellect, *while the button to switch it on is* faith. *Humanity is a wide galaxy, whose center is made up of* intelligence, *while the stars represent* hope. *Humanity is a long, narrow road, made of obstacles and satisfaction, of ups and downs. We know not where it leads to, but we imagine, dream, and so the destination is not the end of the journey, but it's the beginning. Humanity is all these descriptions, one only cannot be chosen, but they can be mixed together.*

Humanity is a great machine, the center of its own galaxy, which follows a road towards its uncertain future. Humanity, one always wonders what it is, who we are? What do we want to obtain? Why? How? When?... Man has always wanted to explore outside himself, forgetting to explore inside himself; sooner or later, I'm absolutely sure, he will find the maturity or simply the courage to do so."

English translation kindly made by Susan Brodar, who is warmly acknowledged.

We also remind about:

"'Tis the times' plague, when madmen lead the blind"

William Shakespeare (1564-1616)

[*King Lear,* Act *IV,* Scene *I, Gloucester*]

Cover Illustration: the cartoon represents Einstein who posed the problem of how he would see the world if he could ride a photon. It was drawn by Lucia G. Gregori (daughter and sister of the authors, and mother of Francesco Spadoni) whom they thank affectionately.

"La pensèe ne doit jamais se soumettre, ni à un dogme, ni à un parti, ni à une passion, ni à un intérêt, ni à une idée"
["Never submit thought to any dogma, or to any party, or to any passion, or to any interest, or to any idea"]
[Dernières pensées, 1913] Jules-Henri Poincaré (1854-1912)

Freedom is a personal feeling, a spiritual conquest. We learn freedom and honesty from our parents and family,[1] from teachers and friends... Everybody's cultural environment dates back to the "discovery" of the "individual" by Socrates (c. 469 BC-399 BC)

Freedom is offended by violence that can be exploited in several ways, even by scientific schools that ridicule every contrasting "non-fashionable" feeling, etc. Science is freedom, not democracy.

We want to acknowledge GPG's wife Rita - and BGG's mother - for her patient understanding while completing this task.

[1] GPG was a teenager, his aunt Carla, the eldest sister of his father, told him: "*some day you will write a book, dedicate it to me*". GPG is glad to fulfill his promise.

Table of Contents

Why this book?

"... quia quarendo, opponendo, disputando et solvendo ac diversas argumentationes et responsiones recitando melius veritas invenitur, et quod a pluribus quaritur, facilius invenitur, ..."
["...in fact, research, controversies, discussion and analysis, and the proposal of different arguments and replies, lead to a better knowledge of truth, and what is investigated by several people is more easily discovered ..."]
William of Ockham (AD 1280 - 1347)

....

What is life?

In this book, it is shown how the entire Universe is dominated by spiral structures - a concept briefly expressed by the term "chirality" - and this applies from the subnuclear world up to galactic superclusters.

As an intermediate stage, there is an endemic tendency for matter to organize into spiral structures typical of organic chemistry - which then evolve into living microorganisms. Life is a universal and endemic phenomenon. Wherever environmental conditions allow, microorganisms evolve into ever larger and more complicated organisms.

That is, the "life" phenomenon exists in the entire Universe, where the sufficiently low temperature allows the formation of atoms and molecules - which can then agglomerate to give rise to forms of life, whose more or less evolved final forms will depend on the environmental availability of sufficient energy sources.

The irreversible flow of time also implies a progressive decay of every structure - living or otherwise - until its total disintegration. However, living structures will have a tendency to - and ability to - remedy damage (antifragility) and to replicate (autopoiesis) - or rather they remedy the inevitable irreversibility of the flow of time with the ability to generate new beings similar to aged ones, in so that the death of these has substitutes which ensure the survival of the species. All this will be explained in detail here.

However, a big preliminary step requires a re-foundation of basic theoretical physics, to show how chirality is intrinsic in all natural phenomena, starting from the world that makes up the interior of atomic nuclei. In this way, it will be possible to get to the bottom of the paradoxes that have plagued – and are plaguing - physics for over a century and more. This will be the leading anthem of this volume. However, the great variety of cases of spiral structures in the Universe, from the micro-world to the macro-world - which is a common property even in the biological or inorganic world - is only briefly mentioned here.

This book is aimed at those *who want to understand the creation that surrounds us* - and who nevertheless feel difficulties - especially when they find themselves faced with arguments burdened by mathematical developments that are not always intuitively understandable.

Mathematics is a marvelous "monument" built by human ingenuity. Mathematics has enabled the incredible successes of engineering and applied science, starting from the period of the Enlightenment until the beginning of the 20th century - and which are still relevant today. In more recent times, classical mathematics has evolved into so-called artificial intelligence (*AI*). Consider, however, that *no mathematical formulation can replace the information provided by Nature's observations*. Indeed, without a concrete and adequate content of observations, every "model" is pure illusion.

The concept was brilliantly expressed by the late Prof. Samuel Warren Carey (1911-2002), authoritative Australian geologist and professor at the *University of Tasmania*: "*garbage in, garbage out*" or if one inserts logical garbage into a mathematical model, the result can't be anything but rubbish. Therefore, you must always enter concrete observations, never fictional contents.

Since the first two decades of the 1900s, with the discoveries of quantum physics and relativity, physics still seems to be stuck in a dead end, full of paradoxes - and it does not seem that any new approach has ever succeeded in solving the mysteries. All this will be much better explained in this book.

However, it must be considered that mathematics is only a "language", which expresses a pre-existing interpretative scheme that is more or less unconscious - and independent from mathematics - based on the philosophy of science, better known as "epistemology". If the basic logical scheme is improper, there is no "language" that can make up for the shortcomings of the approach - just as, if a concept is not correct when expressed in a language, it does not become correct when expressed in a language diverse.

Furthermore, over the last century, science has made incredible leaps and bounds. Science has collected experimental observations that were unimaginable until recently, and which have enabled technological applications that we use every day. This does not change the fact that for a century - or more - basic science, by unanimous belief, has remained sadly burdened by paradoxes, apparently with no way out. Science, therefore, seems to find itself in an impasse, from which it is unable to escape! On the other hand, science must be logical, and we can accept no paradox which in itself is a negation of all logic.

This book deals with the problem from a different point of view. In fact, this book does not assume that mathematics - as a "language" - is essential to explain creation. Rather, it shows how a rethinking of knowledge on correct philosophical (epistemological) foundations can uncover where the logical "bugs" that afflict our knowledge are. *In other words, this book tries to provide a simple and straightforward explanation of the basis of knowledge about the world.*

To this end, in this approach we ignore mathematics as far as possible - which is very useful only at a later stage - and instead discuss *epistemology*. If the reader prefers, we must dust off the classic myth of the "*bon sauvage*" of Jean-Jacques Rousseau (1712-1778), the famous Swiss Enlightenment philosopher, writer, pedagogist, and musician, de *l'Encyclopédie*. That is, we must try to re-establish from scratch our interpretation of what we observed and measured. Our first concern, however, in any case will always be to avoid any logical "bug" that generates paradoxes. This book is a logical journey in this direction.

The mathematical content is here deliberately limited to the minimum required, trying - if and when necessary - to make it easy to understand even for those who are not familiar with it. *In other words, this book is made of simple and realistic reflections - from the most "simple" and essential point of view possible*, where every assertion is motivated by the fact that the assertion seems inevitable, or the "simplest" imaginable, always avoiding paradoxes. This will guide us in a sequence of elementary, obvious - "common sense" - steps which will, however lead us to conclusions that in retrospect will seem surprising and incredible! We will thus have come to terms with the paradoxes that today are a frustrating brake on our knowledge - and this applies both to specialists and to people of culture who have the patience to read us.

This book is composed of small, short, or very short chapters. The reader is recommended - whether he is familiar with scientific texts or not - *to read and meditate on each short chapter, one at a time, in order to ensure that he has fully assimilated the content*. Very often, the reader will in fact have to forcibly "demolish" within himself beliefs that have been rooted in him since his early childhood, and taken for granted and "shared by everyone". Logical "bugs" reside precisely in various such beliefs - and the secret to getting to the bottom of painful logical paradoxes lies precisely in recognizing that many beliefs, now very deeply rooted, have no reason to exist and must be appropriately reconsidered.

Every chapter contains, as far as possible, cross-references to other chapters. Sometimes certain concepts are repeated, and the reader will forgive this.

However, it was all done with a great effort to make understandable many otherwise abstruse or seemingly very bizarre concepts. It was not possible to do better. Therefore, only the reader's goodwill can allow him to understand the meaning of our journey. So let's begin ...

…………………

PS - For those interested, the formulation in the scientific literature is found predominantly in the following papers and references therein.

Gregori, G.P., C.W. Monckton of Brenchley, W. Soon, R. Tattersall, A. D'Amico†, G. Zimatore, V.M. Velasco Herrera, B.A. Leybourne, and Z.A. Oziewicz†, 2022. The golden Ratio, variational principles, cyclic and wave phenomena, and Quanta. In H.M. Colin Garcia, J-de-J. Cruz Guzman, L.H. Kauffman, and H. Makaruk, (eds), *Scientific Legacy of Professor Oziewicz, selected papers from the International Conference "Applied Category Theory Graph-Operated-Logic"* held in honor of Professor Zbigniew Oziewicz in Memoriam (August 25th to 27th, 2021), World Scientific, Series on Knots and Everything; DOI:10.1142/9789811271151.0018: 363-389; http://ncgtjournal.com/additional-resources.html

Gregori, G.P., F.C. Wezel†, L.C. Gregori, B.A. Leybourne, W. Soon, and V. Straser, 2023. Archæology of the concept of "time" in the ancient Western, Eastern, and Far Eastern cultures The foundations of physics, *New Concepts in Global Tectonics, Journal,* 11 (1): 3-34

Gregori, G.P., M.T. Hovland, B.A. Leybourne, S. Pellis, V. Straser, B.G. Gregori, G.M. Gregori, and A.R. Simonelli, 2025. Air-earth currents and a universal "law": filamentary and spiral structures - Repetitiveness, fractality, golden ratio, fine-structure constant, antifragility and "statistics" - The origin of life, *New Concepts in Global Tectonics, Journal,* 3, (1): 106-225

Gregori, G.P., W. Soon, V. Straser, and B.A. Leybourne, 2022a. Golden ratio, fractals, and the arrow of time. Irreversibility vs. reversibility - Space, time, antitime - The foundations of physics. *New Concepts in Global Tectonics, Journal,* 10 (3): 158-201

Gregori, G.P., W. Soon, V. Straser, and B.A. Leybourne, 2022b. The foundations of physics and axiomatics. I - Axioms. *New Concepts in Global Tectonics, Journal,* 10 (3): 202-235

Gregori, G.P., W. Soon, V. Straser, and B.A. Leybourne, 2022c. The foundations of physics and axiomatics. II - Algorithms and equations, and a comparison with Einstein's theory of relativity. *New Concepts in Global Tectonics, Journal,* 10(3): 236-262

Gregori, G.P., W. Soon, V. Straser, and B.A. Leybourne, 2022d. The foundations of physics and axiomatics. III - Superluminal phenomena, mechanisms, matter-antimatter, cosmological implications. *New Concepts in Global Tectonics, Journal*, 10 (3): 263-283

Gregori, G.P., W. Soon, V. Straser, and B.A. Leybourne, 2022e. The foundations of physics and axiomatics. IV - The geometrical representation of symmetries - Parity, emp, and charge (PEC) "meta-multiverse". *New Concepts in Global Tectonics, Journal*, 10 (3): 284-299

Summary

The plan of this volume

We begin our journey like the famous *"bon sauvage"* of Jean-Jacques Rousseau, imagining how we could have seen the world without the *a priori* assumption of Newton's principles - which also have great merits for the development of engineering and applied sciences.

Then we introduce new hypotheses - very timidly - and almost ashamed of going blatantly against the flow. Thus, we build - with incredulity - a "logical castle" of which we ourselves are progressively more and more surprised.

We understand how the perception of "time" is strictly dependent on the organization of the Universe according to the rules of the golden ratio, and how the golden ratio is intrinsically correlated with fractality.

The Universe seems to be in a perpetual search for an equilibrium that it never reaches. If it reached equilibrium, everything would end, and the entire Universe would be a still swamp where nothing would happen. There would be no life, there would be no perception of "time", both by man and by any living form.

All this strictly implies irreversibility - in stark contrast to the Galilean teaching, which prescribes isolating individual natural subsets and looking for chains of cause and effect, on the assumption that every observed state is one of complete equilibrium. The Galilean criterion has immense merit for engineering and applied sciences, but it is based on an assumption that is intrinsically non-physical, as it is opposed to irreversibility.

We find out how it seems that the existence of antimatter can be explained only by supposing that the truly absolute world is the timeless β-world, which occasionally generates two symmetrical worlds - which in mathematical language are complex conjugate of each other – i.e. the world α which is our world with matter, time and entropy, and the world γ with antimatter, antitime, and antientropy.

Another important concept is that it is necessary to distinguish between the micro-cosmos and the "statistical" treatment of phenomena. This holds in everything that we can or cannot objectively observe due to the empirical constraint.

We discover how it is essential to acknowledge that Nature is organized in monads, i.e., in units confined in space and time on different scales, the knowledge of which is very different for the monads that we can know from the inside, and those that we can only know from the outside. The latter are only known to us through sub-monads that the monads expel, as with a "sling-shot".

Furthermore, it seems obvious to conclude that the laws of "statistics" - which are valid in our macro-world of lotteries, ballot exit polls, etc. - cannot apply inside other monads. Thus, the much celebrated Schrödinger equation is none other than the correct "statistical" treatment of the atom monad. Furthermore, as such, the same equation cannot be applied to monads larger or smaller than the atom, such as to the monad atomic nucleus.

Finally, we enter the world of life that thrives in the world α, and of the antilife that thrives in the world γ, having however to introduce a new concept of "statistics", which is taken up from the sciences of the financial world, and we discover the very important role of *antifragility*. As a subsequent extrapolation, we realize the fundamental role of autopoiesis. That is, considering that irreversibility - i.e., the flow of time - means that everything that exists deteriorates, "ages", and finally "dies", the monads that exist have exploited the capability to generate new monads that replace those that "die". In fact, in consideration of irreversibility, nothing can be eternal, but everything must regenerate in order to replace with new systems those which, due to the increase in entropy, can no longer exist and must die.

This statement also applies to galaxies and antigalaxies. In fact, to our ever greater surprise, we sense that it appears more than reasonable to suppose that life is an endemic phenomenon in the Universe - and we consider some experimental evidences - both in the depths of the Earth's oceans and in the morphological characteristics of bodies in the Solar System – that are consistent with this fascinating hypothesis. In fact, the life forms, which ought to exist in the different bodies scattered across the Universe, correspond to the objective possibilities of survival and reproduction - in order to allow autopoiesis.

We can thus imagine the existence of galaxies of matter, and of antimatter galaxies – which is a hypothesis originally formulated by Hannes Olof Gösta Alfvén (1908–1995), Swedish physicist, Nobel Prize winner for Physics in 1970. Galaxies of matter and antimatter – which are two objects that are perfectly mirror images of each other - sometimes annihilate upon entering the timeless β world - and in this way, the entropy (which has grown disproportionately in galaxies) and the antientropy (which has similarly grown in antigalaxies) become zero.

An enormous phenomenon of materialization from the timeless world β will then be able to generate new galaxies and - symmetrically - new antigalaxies, which are complex conjugate to each other, thus completing a cycle that can be repeated indefinitely.

Finally, we will show how in the atemporal world β it is not possible to define an extension, given that the concept of 3D space applies only in the world α or in the world γ. Therefore, the big-bang model, characterized by a huge explosion from a central point, makes no sense, as does the big-crunch model, characterized by a huge implosion at a single point. Big-bang events and big-crunch events are measurable in a 3D space in the α world or the γ world, but not in the timeless β world. Therefore, the galaxies and antigalaxies that we observe with our telescopes are generated at sites that, for observers in the α world or the γ world, are randomly located in 3D space. Cosmological models must therefore be fundamentally rethought.

While all this study was being perfected in spelling and orthographic details, the prestigious journal *Proceedings of the National Academy of Sciences* published a paper by Michael L. Wong and collaborators (of the *Earth and Planets Laboratory, Carnegie Institution for Science,* Washington, DC) where the "*Law of the increase in functional information*" is formulated, which is a hypothesis that in some way is vaguely connected to the unifying vision of the perspective of our path, although starting from the laws of motion, gravity, electromagnetism, and thermodynamics.

In contrast, our approach ignores any previous concept, with the exception of Feynman graphs. Indeed, we greatly broaden the perspective by explicitly including the entire Universe from the subnuclear scale up to galactic superclusters. Furthermore, we resolve the various paradoxes that have plagued theoretical physics for a century - while we also include phenomena related to life in the Universe. The "*Law of the increase in functional information*" is therefore generalized here to the principle that can be called the "*Law of antifragility in the Universe*", where the search for harmony is linked to the tendency towards equilibrium, in accordance with the order required by the golden mean.

Enjoy your reading.

"There are more things in heaven and earth, Horatio,
Than are dreamt of in your philosophy"
(*Hamlet*, Act I, Scene, IV, 166-167)

Chapter 1

The Empirical Constraint

Dirac, Fermi and Feynman on the fundamental role of intuition

"I understand what an equation means if I have a way of figuring out the characteristics of its solution without actually solving it."

Paul Adrien Maurice Dirac (1902-1984)

Fermi liked to joke whenever his colleagues wondered about his skill in giving, intuitively, the correct physical interpretation of some observations. He claimed that he used "c.i.f.", which are the initials for the Italian words "*con intuito fenomenale*", i.e., "*with intuition of phenomena*", the joke being in the fact that the adjective "*fenomenale*" has also a different meaning, being equivalent to "extraordinary" or "amazing".

"The physicist needs a facility in looking at problems from several points of view … [A] *physical understanding is a completely unmathematical, imprecise, and inexact thing, but absolutely necessary for a physicist. …"*

Richard Phillips Feynman (1918-1988)

……………………………………………

Let's consider a beautiful garden with a nice pond inside it. Let us consider how this enchanting place is perceived by different living beings - such as an earthworm, or an insect that lives underground, or an ant that laboriously walks among the blades of grass, or a bee that assiduously searches for floral pollen, or a turtle, or a lizard looking for a stone on which to enjoy the sunlight, or a mouse, or a cat, or a dog, or a horse, or a cattle, or a sparrow, or a swallow, or a hawk, or an eagle ... finally how the garden is perceived by a fish that lives in the pond. Man is no exception to all other beings.

The perception that man has depends on his senses, on his physical size, and on his lifespan. Man considers "large" everything that has dimensions larger than his own (such as an elephant, a giraffe, a cattle, a horse, a whale, a shark, etc.), and considers "small" everything that has a smaller size to its own (a cat, a dog, a bee, an ant, etc.).

If man had been endowed by Mother Nature with gills like fish, or with lungs but with the freediving faculties of dolphins or whales, he would have explored and occupied - instead of the emerged lands - the sea and ocean floors which today are a very large unexplored part of the Earth.

The senses of man are fundamental. The knowledge that man can have of the environment depends first and foremost on how much he can perceive empirically. Hence, the meaning of "*empirical constraint*", from which it is not possible to escape. It is up to man to "logically" organize his limited empirical information, to find what in Nature seems to follow some "*laws*" that allow us to "predict" how the phenomena occur - all to the benefit of man's survival and quality of life. This is the definition of "*natural science*". The first step - for an epistemological analysis - is therefore to become aware of the implications and consequences of the *empirical constraint.*

To this end, we preliminarily distinguish what we perceive visually from the no less important perception of time.

Man perceives what he sees, i.e., what emits light (photons, i.e., grains or "quanta" of light), and light manifests itself as a wave of appropriate wavelength (i.e., color). Man perceives light in a narrow band of wavelengths, which corresponds to the colors of the rainbow – red light has a longer wavelength, violet light has a shorter wavelength, and all others are in between colors. But there is a very broad "spectrum" of light frequencies that man does not see. Light with a wavelength longer than the wavelength of red light is called *infrared* light, while light with a wavelength shorter than the wavelength of violet light is called *ultraviolet* light.

For instance, the sky is very bright in infrared light. Some animals see infrared light, humans do not. Man thus resorts to infrared viewers - now common in homes for remote surveillance - and infrared photographs or television footage are well known, as well as the obvious military applications. That is, man can use his senses, together with appropriate instruments that "extend" his perceptive abilities.

However, if what should be observed does not emit any type of light, man is unable to detect its very existence. It appears that the vast percentage of the matter in the Universe is of this type. We cannot "see" this matter even with sophisticated equipment. This is the so-called "*dark matter*", whose existence can only be indirectly perceived through its gravitational effects. These topics will be clearer as we continue our logical journey. For now, these simple preliminary comments will suffice.

Anticipating concepts that are better clarified later, we ask ourselves how the emission of light is possible? The emission of light – i.e., of photons from an atom - presupposes that atomic nuclei have a "shell" of electrons on appropriate orbits - a bit like a miniature Solar System. When an electron is thrown from an inner orbit to an outer one, the atom emits a photon – and the opposite happens when the electron is thrown from an outer orbit to an inner one, and the atom absorbs a photon.

However, when matter is extremely hot - much hotter than on the visible surface of a star - then *all* the electrons of the "shell" are "torn" from their respective atomic orbits, leaving the atomic nuclei "naked" - a bit like a Sun or star without any planets. No light emission can then take place, since the essential instrument for the release of light is missing - unless we consider radioactive emissions etc., but that is another matter, and we will return to this later in our journey (see Chapter 15).

In summary, man knows Nature only as long as he is able to observe some signal sent by a component of the natural system. That is, man's capacity for observation strictly depends on this essential aspect of the "*empirical constraint*". Any phenomenon that is not observable empirically *cannot be part of human science*. One can only rely on considerations of the imagination. Such considerations are called "*speculative*" and are valid only as long as they are confirmed by what has been observed - otherwise they must be rejected and replaced by new speculations that are in better agreement with the available concrete observations.

This is a difficult and also frustrating procedure, and every result must always be questioned as soon as new experimental observations become available. Enrico Fermi (1901-1954) said that when an instrument is operated, there are two possibilities: either the results confirm the hypotheses and then a *measurement* has been carried out, or when the result is contrary to the hypotheses, then a *discovery* has been made. In other words, the engineer must make measurements, the scientist must make discoveries.

Furthermore, when there are several hypotheses or explanations, they must all be taken into account as possible guesses - and one must look for observations that can discriminate between any doubts. In any case, when in doubt, an excellent criterion is that of "maximum economy", also called "*Ockham's razor*". That is, you need to choose the simplest solution with less speculative content, so as to reduce the risk of subjective suggestion.

This is the meaning of the difficult progress of science – and this is the meaning of the journey proposed in this book. It goes without saying that "para-normal" or "mystical" or "superstition" explanations are outside of "science". Science is logic, and as such, it can accept no paradox. If we find paradoxes, there are "bugs" in our basic setting. We have to change. This is the objective of our journey.

In addition to visual perception, another *fundamental* aspect - which will be the subject of repeated reminders later in our journey - concerns *the perception and measurement of time*. What "time" is has been the subject of heated debate since prehistoric times, and will be discussed at length *passim* in all our journey. For now, we accept the fact that time exists - and everyone perceives it, just as every living being in the animal or plant world perceives it, since, as we will see, the perception of time is essential for survival itself. Perhaps man retains memory of it, unlike other life forms. However, a living organism, even a very basic one, which does not perceive time, cannot know when to feed, sleep, procreate, etc. Or he doesn't survive.

How do we measure time? We rely on some cyclical phenomenon that we hope should repeat regularly with a constant repetition period: for simplicity, we call this cyclical phenomenon the "*reference oscillator*". It should be emphasized that without a reference oscillator, we cannot measure time - and the temporal resolution of phenomena in time measurement cannot be more detailed than the cycle-period length of the reference oscillator.

This aspect is important, and it is one facet of the empirical constraint. We can only measure time intervals larger than the period of the reference oscillator period.

The most basic reference oscillator is the pulsation of our heart, which is also very unreliable, considering the period variations connected to our physical effort, our emotions, state of health, etc.

Galileo understood the usefulness of a pendulum, and it was a "lightening" discovery. Since then, pendulum clocks have become an indispensable instrument for measuring time.

Previous devices - since ancient times - were only sundials (on vertical or horizontal surfaces, see below). Note that hourglasses measure an interval of time starting from the moment the hourglass is turned upside down, i.e., hourglasses measure a duration, certainly not the time of day.

Sundials, however, require a fixed surface (horizontal or vertical), suitably exposed to sunlight. Sundials require clear skies. They work only during the day. They require adequate calibration. They are indispensable for measuring the progress of the seasons. Their information was essential even in the most distant antiquity, when man became sedentary in the Neolithic, and for survival it was thus indispensable to know when to sow and harvest, to migrate livestock, under penalty of very heavy famines, diseases, etc. (this extremely interesting aspect of prehistory is studied by *archæoastronomy*).

Mechanical devices, alternatives to sundials, based on mechanical oscillators (i.e., pendulums or balance wheels) led to the development of the famous historic clocks that still grace the squares of numerous European cities. It should be noted, however, that in general, a sundial - less visible - is generally placed almost surreptitiously next to the sumptuous mechanical clock. In fact, the sundial was used to check that the clock had no drift – i.e., that it did not advance solar time or delay it. In those days, there was no exact time service, which is now ensured by satellite systems that control portable phones, PCs, etc. Older people remember that after the Second World War, the radio emitted every day in the early morning a "time signal" on which every listener could correct their mechanical clock.

The most accurate reference oscillators today are atomic. Furthermore, thanks to particular and very delicate electronic devices, it is possible to reach even finer temporal resolutions, but only for very special experiments. However, there always remains a limit of resolution beyond which you cannot go.

Therefore, just as man calls "large" or "small" everything that refers to his size, man measures what is "short" or "long" by referring it to the average duration of his life - rounding up, we say that the reference is a period of one hundred years. But this can be misleading.

In fact, we know well that some animals have an average lifespan much shorter than that of humans, while – e.g. - tortoises are typically much longer-lived. The formation and destruction of continents occur on the scale of *many tens of millions* of years (*Ma* = *Mega* years), or *a few hundred million years* (of *Ma*), while the formation of the Solar System dates back to approximately 4.6 *billion* (that is, *Ga* = *Giga* years) of years ago. The evolutionary cycle of a star is of the order of almost 100 Ga, and that of galaxies is unknown, but much, much longer still. All this will be a little clearer (if possible) at the end of our path.

In any case, man is not the center of the Universe, and it makes no sense to relate everything to the average duration of man's life.

With this, we have completed a brief introduction to the empirical constraint, in its essential implications, both with regard to what we perceive with our senses or with our instruments, and with regard to the perception and measurement of the flow of time. We will necessarily return to these topics at the appropriate time: we are only at the beginning of our journey!

The measurement of latitude and of longitude

Only as a curiosity – aimed to "enlighten" our description of natural phenomena, and in order to show how this knowledge can have unexpected, even tragic, consequences - may we be allowed a small, and in some respects terrible, historical excursion. It concerns the measurement of longitudes on the globe, and the misunderstandings which, e.g., have led, as we try to briefly explain here, to tragedies such as those which inspired the well-known 1986 movie "*The Mission*" by Roland Joffé, with a famous music by Ennio Morricone.

Let's consider how latitude and longitude can be measured on the globe.

The measurement of latitude is comparatively easy. In principle, it is sufficient to measure the height above the horizon of the Pole Star (Fig. 1.1 is easily understandable even on the basis of a few basic geometric notions, but the understanding is not essential here).

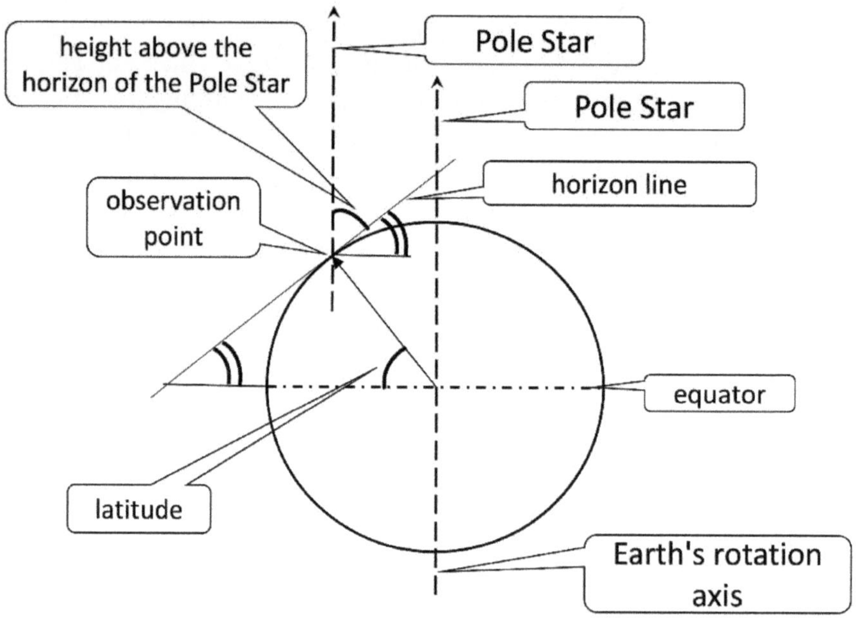

Fig. 1.1. The figure shows a section of the Earth through the observation point and through the geographical poles. The latitude and the altitude above the horizon of the Pole Star at the observation point are equal, as can be shown by elementary geometry. But understanding this is not essential for what follows.

Alternatively, you can measure the maximum height above the horizon of the Sun every day at a specific site whose latitude you want to know. If you know the progressive day (date) over the course of the solar year, you can deduce the latitude of the observation point from a predefined table that indicates the maximum height of the Sun at local noon for every latitude.

In prehistoric times, this default table was not available. However, it was easily possible to establish whether two locations - even very distant from each other - were at the same latitude. That is, in this way, it was not possible to calculate the latitude, while it could be ascertained whether the latitude of two locations was the same or not. However, the North Star made it possible to measure the correct latitude of every location (Fig. 1.1). But above all, with this type of observation of the shadow of the Sun's rays, it was possible to recognize the season, and establish when to sow and harvest, or undertake transhumances, etc.

For carrying out these measures, an esplanade was necessary of beaten soil - kept clean of grass, and located above any obstacle that provided shade. In the center of the esplanade, a pole was raised vertically, let's say at least $5-6\,meters$ high. It was thus possible to establish whether the latitude of two locations was identical, with an error of the order of $2-3\,minutes$ (approximately 5 km). In fact, the minimum length of the shadow at local noon must be identical in the two locations of the same latitude.

This method was applied in the pre-classical world and in the classical Greek world, and was also the inspiring principle of the Sumerian *ziggurats*, or of Monte d'Accoddi in Sardinia (Fig. 1.2 and 1.3). But the examples are almost endless.

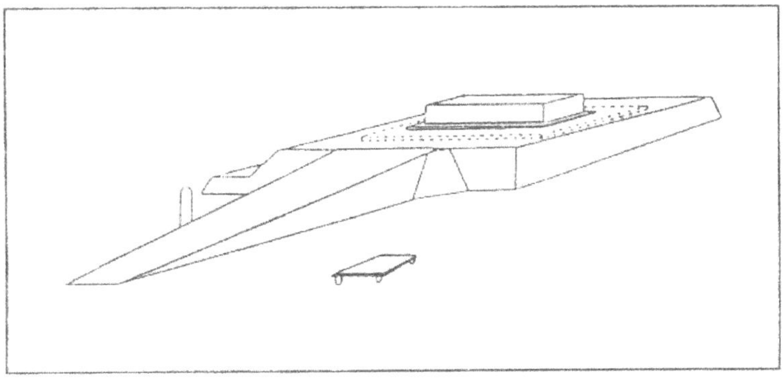

Fig. 1.2 – Mount Accoddi is located in Sardinia, not far from Sassari. A ramp made it possible to drag large poles, many meters long, up to an open summit esplanade. The building shown here at the top of the esplanade is an addition, made when the cycle of the seasons had already been established, or maybe imagined by archæologists. See Fig. 1.3.

Incidentally, the ancient Egyptians used a much more accurate and cumbersome tool. For instance, the Karnak temple consisted of a sequence of courtyards accessible only by a gate. The gates were all aligned, with the central axis oriented in an East/West direction so that a ray of sunlight could enter and reach the final temple only on the winter solstice. Perhaps this was to celebrate and welcome the rising Sun. This astronomical instrument was part of the difficult study carried out by the Egyptian astronomers/priests, who were concerned with predicting the floods of the Nile, and therefore studied and compared the four different calendars based, respectively, on the stars, on the Sun, on the Moon, and on Venus. Their astronomical wisdom made the ancient Egyptians the most famous true astronomers (not astrologers) of antiquity.

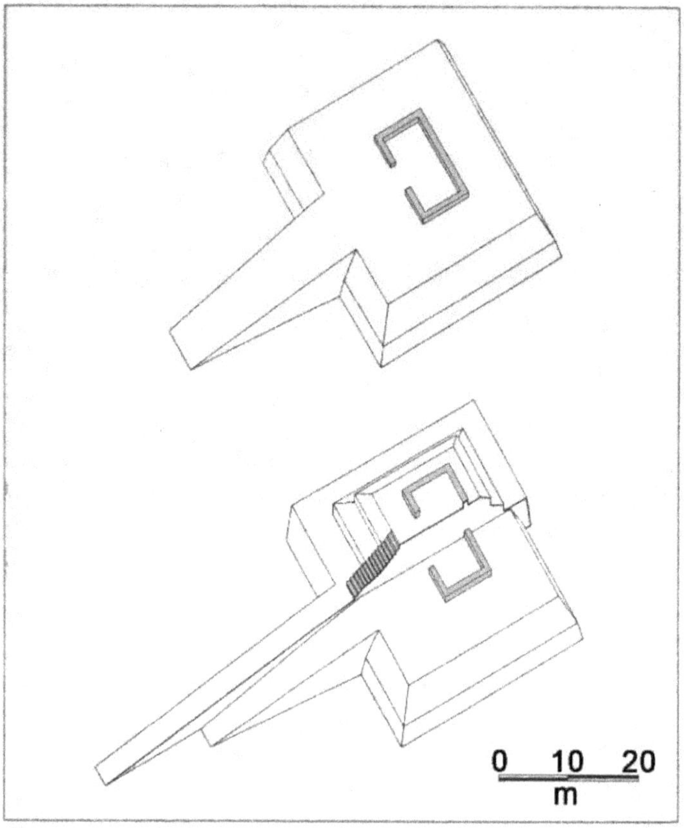

Fig. 1.3 – Mount Accoddi was uplifted and enlarged, evidently because some new constructions covered the horizon. The building on the summit esplanade was either the result of later additions or imagined by the archæologists. On the summit esplanade of the smaller "Mount" clear traces of fires were found, due either to herbicide actions or to lightning that struck the vertical pole. The construction of the ramps involved the movement of approximately 400 m^3 for the smaller ramp, and approximately $2000 - 2500\ m^3$ for its enlargement. Considering the small population density in Sardinia in Nuragic times, and considering the very limited technical means, it is hardly credible that so much effort was made except out of strict necessity. Figs 1.2 and 1.3 are taken from studies carried out by the late Prof. Santo Tiné of the *University of Genoa*, whom we thank.

Conversely, the measurement of longitude was much more difficult. The only possible way was to measure the local time at every given site - based on the elevation of the Sun above the horizon at a certain instant - while knowing the exactly simultaneous local time at a point on the Earth, a point which was considered as reference. Today, there are very accurate satellite services that assist airplanes and ships. But the astronomical observations were applied until artificial satellites became available. Owing to this reason, it was essential for every ship or plane to have an officer in charge of the route, etc.

Historically, the reference point was chosen at the *Greenwich Observatory*, a suburb of London, which was then the most important maritime port in the world. Therefore, it was then strictly essential to have, on board every vessel, a clock that maintained - with great stability and reliability - the Greenwich Mean Time information. However, the pendulum invented by Galileo suffered terribly from the rolling and pitching of the vessel. It just couldn't work!

Galileo had also proposed to use the observation of the satellites of Jupiter (Io, Europa, Ganymede, and Callisto) that he had discovered with the "telescope" (which he called "*perspicillum*"). But the optical technology of the time did not appear to be practically reliable for application across all ships.

The problem was certainly very serious and heartfelt, and the best minds of the time tried their hand at it, until John Harrison (1693-1776), an English watchmaker, invented the "marine chronometer" - but this invention only occurred in 1761, and finally allowed precisely determine longitude on the high seas. The device was based on a system of springs and shock absorbers, such as swinging bars around a horizontal axis, etc., which made it indifferent to oscillations and vibrations. Harrison built only a few examples - today in the *Greenwich Observatory Museum* - more and more perfect, but he had only partial recognition of it, until Captain James Cook, after using it during the three years of navigation of his second voyage (1772-1775), expressed enthusiastically about it.

Before Harrison's "marine chronometer", longitude was estimated with notable errors, as can be seen from the first historical geographic maps of the Americas.

At the end of the 15th century, there was a very strong competition between the two maritime superpowers of the time, Spain and Portugal, until they repeatedly asked the Pope to mediate. These were the times of Columbus' voyages to the New World. Various Papal Bulls were issued in order to carry out mediation action. In 1494, in an attempt to find a peaceful compromise between the two parties, the Treaty of Tordesillas was signed, which divided the spheres of influence of the two maritime powers on the basis of longitude. It thus became crucial to define the "Tordesillas line", whether to extend it only to the Atlantic Ocean, or to the entire globe, etc.

At the beginning of the 16th century, Spanish and Portuguese geographers identified a point at the mouth of the Amazon River as the intersection point of the Tordesillas line with the American continent. The "diplomatic" story is tragically very intricate, lasting until the discovery of Harrison's "marine chronometer" - with atrocious episodes as remembered in the movie "*The Mission*". That is, already flourishing and consolidated missions were destroyed, massacred, reduced to slavery, just because their longitude did not correspond to the sphere of influence foreseen by the "Tordesillas line". Atrocity has always resided in the human soul.

In conclusion, the empirical constraint must be taken into account when evaluating any "physical law". Indeed, what are the real bases of every proposed "law"? What are the objectively reliable bases, and to what extent does our presumed knowledge derive from speculation, or from an arbitrary or convenient interpretation?

These are the questions we try to answer on our journey.

Chapter 2

The Illusion of Continuity

On the definition of "*trumpets*," and the role of the "idea"

*"So bene che, per non essere io litterato, che alcuno presuntuoso gli parrà ragionevolmente potermi biasimare coll'allegare io essere omo senza lettere. Gente stolta! … Diranno che, per non avere io lettere, non potere ben dire quello di che voglio trattare. Or non sanno questi che le mie cose son più da essere tratte dalla sperienzia, che d'altrui parola; la quale fu maestra di chi bene scrisse, e così per maestra la piglio e quella in tutti i casi allegherò. … Molti mi crederanno ragionevolmente potere riprendere allegando le mie prove essere contro l'alturità d'alquanti omini di gran reverenza appresso de' loro inesperti iudizi, non considerando le mie cose essere nate sotto la semplice e mera sperienza, la quale é maestra vera. (*Atl 119 v.a*). Se bene, come loro, non sapessi allegare gli altori, molto maggiore e più degna cosa allegere allegando la sperienzia, maestra dei loro maestri. Costoro vanno schonfiati e pomposi, vestiti ed ornati, non delle loro, ma delle altrui fatiche; e le mie a me medesimo non concedano; e se me inventore disprezzano, quanto maggiormente loro, non inventori, ma trombetti e recitatori delle altrui opere, potranno essere biasimati … È da essere giudicati e non altrementi stimati li omini inventori e 'nterpetri tra la natura e gli omini – a comparazione de' recitatori e trombetti delle altrui opere … Gente poco obbrigate alla natura, perché son solo d'accidental vestiti, e sanza il quale potrei accompagnarli infra li armenti delle bestie. (*Atl 117 r.b*)"*

[*"I am well aware of the fact that, since I am not a literary man, some presumptuous guy will believe reasonable claiming that I can be blamed because I have no classic background. Stupid people! ...They will claim that, since I have no classic learning, I cannot well state what I want to discuss. These people do not know that my items are taken from experience* [i.e. from observations] *better than from others' opinion; experience was the teacher of all people who correctly wrote, and thus I take it like my teacher, and in all cases I will refer to it… Several people will believe reasonable that I can be criticized by arguing that my proofs are against the authority either of several other people of 'great reverence' or of their inexpert judgments. But they will not consider that my inferences are based on simple and mere experience, which is the true teacher* (Atl 119 v.a). *Even in the case that, unlike them, I cannot quote authors, I will refer in terms of a greater and more worthy approach by quoting experience, which is the teacher of their teachers. These guys go on inflated and with great pomp, dressed and adorned with the achievements by others, rather than by their inferences; and if they do not acknowledge my results; and if they do not consider me as an inventor, even much more than this they shall have to be blamed as they are not inventors, rather trumpets and actors who perform the results of others ...One has to appreciate and definitely estimate all people who are inventors and who interpret nature for other men - compared to actors and trumpets of the achievements by others ...They are people who are poorly obliged to nature, because they are dressed only with accidental items* [i.e. with conventional, generally agreed, beliefs], *and owing to such deficiency I could well put them among herds of beasts. (Atl 117 r.b)"*] [Our English translation]

[Leonardo da Vinci (1452-1519)]

"La pensée n'est qu'un éclair au milieu d'une longue nuit. Mais c'est cet èclair qui est tout."

[*"Thought is like a lightning in the middle of a long night. But this lightning is everything."*]

[Jules-Henri Poincarè (1854-1912)]

....

We need to start with a simple warning for every reader who is not accustomed to looking at a graph. There is nothing difficult about it.

It must be considered that a table of numbers seems to everyone - even to the specialist - to be a dry list with no particular meaning. However, when these numbers are plotted in graphic form, the eye - which is a powerful natural computer - immediately and instinctively perceives regularities, patterns, etc. In fact, we must consider that the data "speak", as they generally refer to measurements carried out in Nature - and this is a way of making Mother Nature "speak", and of understanding its messages. So every reader should, if necessary, make a small effort to understand the "language" associated with a plot. All of this has nothing in common with mathematical language at all. Rather, this is part of the underlying, intuitive language of understanding Nature.

For clarity, we have here invented a set of *50* numbers that can represent any physical quantity that we will call Y (such as a mass, or an energy, or a pressure, a macro-economic data, or something else, or even an index such as a temperature, or something else). Every number is hypothetically measured in *50* successive moments of time that we call X (for simplicity, we assume at regular time intervals). Of course, the concept also applies when X is not an instant in time - and when the values of X are not spaced at irregular distances. We have also invented, arbitrarily, the uncertainty of every measurement (called "*measurement error*"). The error denotes the probable fluctuation margin of the measured value Y due to the limited precision of our measuring instrument. The result is displayed in Fig. 2.1. Another form of representing the same (hypothetical) series of measurements is that of a bar graph (Fig. 2.2).

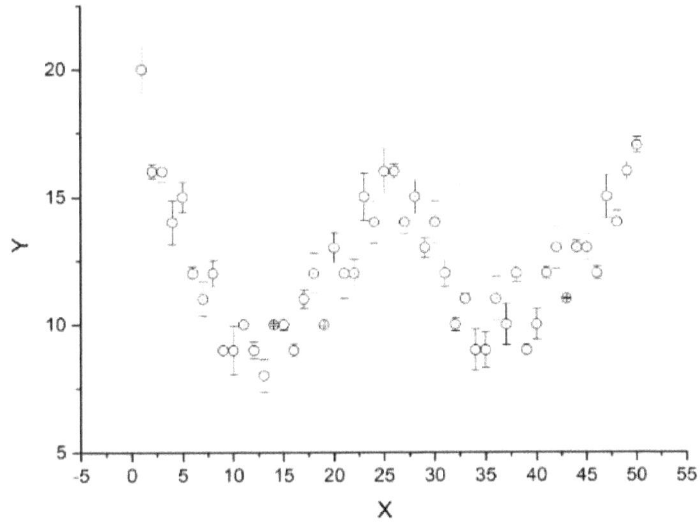

Fig. 2.1 – Plot of *50* hypothetically measured values of Y as a function of the value X (which may be, e.g., an instant or something else, either at regular or non-regular distances). The vertical bars indicate the measurement error of every specific datum.

It is widespread practice to interpolate the measured Y values with a line. We show this in Fig. 2.3 - where the measured points with relative measurement errors are indicated (as in Fig. 2.1), and the *interpolating curve* is superimposed on all given points. A complicated discipline shows how to interpolate the data, taking into account the relevant measurement errors, and on the basis of "reasonable" hypotheses. This discipline is not of concern here. The curve used here - in words well known to a first-year middle school student - is a simple polynomial of degree 9 of X, or of the type $Y(X) = a_0 + a_1X + a_2X^2 + a_3X^3 + a_4X^4 + a_5X^5 + a_6X^6 + a_7X^7 + a_8X^8 + a_9X^9$, where the constants a_0, a_1, a_2, ... a_9 have been appropriately calculated.

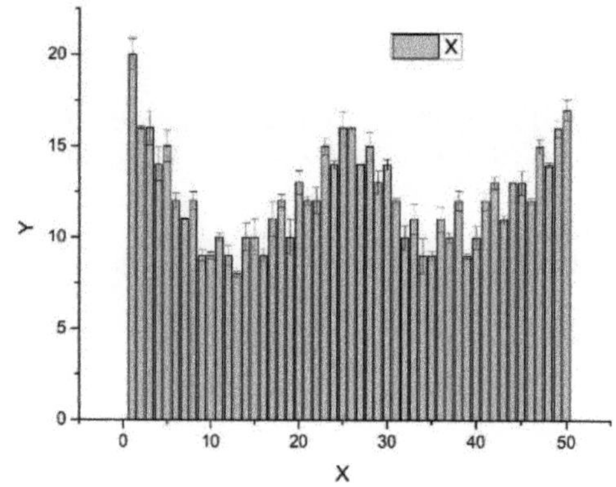

Fig. 2.2 – Bar graph of the *50* hypothetically measured values of Y already shown in Fig. 2.1. That is, Figs 2.1 and 2.2 are identical, changing only the way of plotting and displaying the identical physical contents.

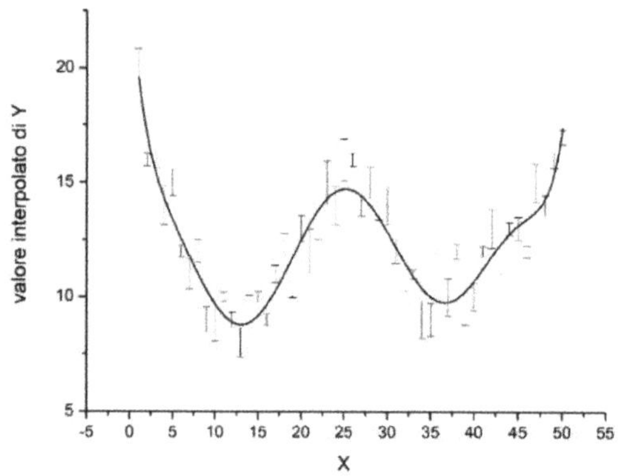

Fig. 2.3 – The same *50* hypothetically measured values of Y shown in Fig. 2.1, interpolated with a line.

For simplicity, it is customary in the physical discussion of the results to consider only the interpolating curve, i.e., to consider only Fig. 2.4, by effectively forgetting the series of measurements actually performed and shown in the previous figures. Without realizing it, using only "obvious" common sense, we have thus introduced a concept that to us seems simple, almost innate, common - but which in fact is devastating for our understanding of Nature: we have replaced a discrete series (i.e., made up of a sequence of a finite number of data) with a "continuous" line and we have deluded ourselves that this is the best way to synthesize our knowledge of reality. This is what we can call the *illusion of continuity*.

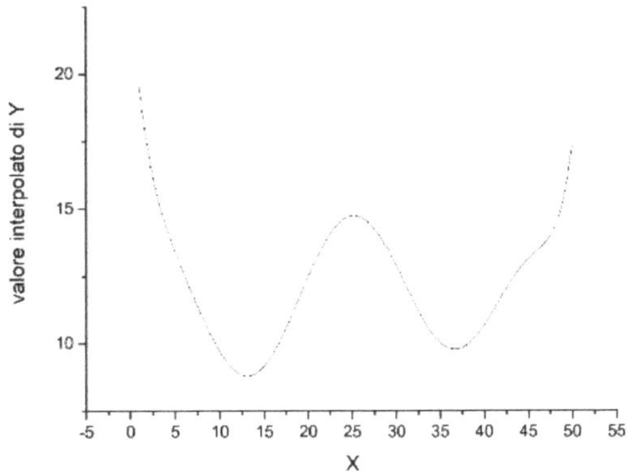

Fig. 2.4 – The same Fig. 2.3 when the measurements actually performed are not plotted.

How can we define "*continuity*"? Let's consider two "very close" points of the curve in Fig. 2.4. We suppose that - however close the two points are - there always exists at least one other point between the two chosen points. That is, we say that the line is "*continuous.*" With this, we also introduced the important concept of "*infinitesimal*": a quantity is said to be infinitesimal when - considering a "very small" value such as the distance between two points of a "*continuous*" curve - we suppose that a value always exists that is smaller than the previous one.

The concept is more common than you might think. Consider, e.g., in everyday elementary geometry, the concept of "point", "line", "surface", etc. They are all the result of an abstraction, because "point", "line", "surface", etc., do not exist in objective reality.

For instance, a planet is considered a "point" - with its own mass - that moves around the Sun, while - in fact - every planet is an object of considerable dimensions.

Similarly, we consider a "line", with zero ideal section, a piece of pasta spaghetti, or the trace of a pen on a sheet, while in fact they have a cross-section or thickness much larger than zero.

Similarly, a thin sheet of paper - or of pasta for preparing baked lasagna - is considered of negligible thickness, i.e., an ideal surface that is flat or not, while in reality it has a well-defined thickness, etc. Consider, e.g., also the Milky Way. The Milky Way is a sort of "pizza" containing myriads of stars, and has a diameter of about $300{,}000\ AL$ (light years). That is, the light speed, which - as is well measured and established, is $\sim 300{,}000\ km/sec$ - takes approximately $300{,}000\ years$ to travel the entire diameter of the "pizza". This "pizza", with the exception of a relatively small inner core, has a thickness of the order of only about $\sim 400\ AL$. That is, the thickness/diameter ratio of this "pizza" is of the order of $400/300{,}000$, which is comparable to the thickness/size ratio of a sheet of paper. So we can consider the Milky Way approximately a "surface".

In other words, for "simplicity", depending on what we want to calculate, we make an abstraction and consider a given real object "approximately" similar to an abstract concept. The same thing goes for every parameter, e.g., let's consider two objects that slide *ideally* without friction, or a *perfectly* elastic object, or a *perfectly* non-deformable body, or an *ideal* gas, or an *ideal* liquid, etc.

Even the concept of "*infinitesimal*" quantity is the fruit of an abstraction, made for our convenience or for "*simplicity*".

Similarly, the concept of "*infinity*" is the fruit of an abstraction. A quantity is said to be "*infinite*", when - considered a "*very large*" quantity - we assume that an even larger quantity always exists. For example, by having increasingly powerful telescopes placed outside the Earth's atmosphere, myriads of previously unknown galaxies are being discovered - and we have certainly not seen the boundaries of the Universe. By this, it is customary to say that the Universe is "*infinite*". In reality, this statement is only a way of confessing our inability to see beyond a given boundary – and this boundary, if it exists, is essentially connected to the empirical constraint. That is, with our measurement instruments, we can explore the Universe up to a certain boundary. Then, we sail in ignorance…. The empirical constraint always imposes its rules.

In summary, "*infinitesimal*" and "*infinity*" are concepts derived - such as the *illusion of continuity* - from convenient abstractions which - due to the empirical constraint - cannot have an objective comparison with the reality observed in Nature. It is therefore very probable that they are fictions created by our speculations. They could therefore be completely misleading.

Extrapolations

Obviously, when we refer to the curve in Fig. 2.4, there is a great temptation to extrapolate the curve - which has a known analytical form (in this case, a polynomial) - either backwards or forwards. For instance, if X expresses a time instant, and if the value Y is a parameter of climatic interest, we can try to reconstruct the value of Y in a distant past, or in a distant future. In reality, it must always be considered that any extrapolation into the past or future is the result of arbitrariness, and *it is essential to seek observational evidence that shows to what extent the extrapolation makes sense and is confirmed by reality*. Without such feedback, every conclusion is the result of an *arbitrary overestimation of the meaning of mathematics compared to true natural phenomena.*

Concerning the distant past, indirect evidence – even though often very tenuous - is sought from geological or palæontological evidence. As far as the future is concerned, meteorologists constantly and conscientiously compare their predictions with what actually happens. Thus, meteorologists know that, today, forecasts - despite some uncertainty - are on average reliable only for about 5 subsequent days or a little more, while they are unreliable in the longer term. As for the case of longer-term "climate" forecasts, it goes without saying that they are in fact questionable at best.

Rational numbers and irrational numbers

Another important concept - much more than a simple "curiosity" of interest only to mathematicians - concerns the digits that, after the comma, express the decimals of a given number. Suppose we write all the decimals of a given number, even if there are a lot of them. For the continuation of our journey, it is very important to foresee two cases as follows.

A number category has a finite number of decimals. The decimals can also be many, but after that certain number, every further decimal is strictly zero. These numbers are called "*rational*" and it is easily shown that they can always be expressed as the ratio p/q between two integers p and q. All numbers that can be expressed by the ratio p/q between two integers p and q, including the numbers with an infinite number of periodically repeating decimals, are "*rational*".

However, there is also an even larger category of numbers that necessarily have an unlimited number of decimals. That is, they cannot be written entirely with all their decimals, and they cannot be expressed as a ratio of integers of the type p/q. These numbers are called "*irrational.*" The best known cases are, e.g., π or even $\sqrt{2}$, etc.

The distinction between *rational numbers* and *irrational numbers* will be recalled below - being fundamental to understanding natural phenomena, and in particular to understanding the meaning of the flow of "time".

But why do we say that "continuity" doesn't exist in Nature?

In fact, for millennia, it has always been believed that Nature is continuous, until we came across a disturbing mystery or paradox. The historical importance of this discovery is fundamental - and therefore deserves to be described here in some detail. It should be in the baggage of every cultured person.

We need to introduce the case of an object (e.g., a small sphere) covered in carbon black. On the basis of an abstraction (such as those above), we consider an *ideal* object - which we will call a "*black body*" - defined as an object characterized by the fact that it absorbs integrally any light radiation it receives of any wavelength. Then, it promptly re-emits the entire received radiation as new radiation that is spread over different wavelengths, depending on the temperature of that ideal "*black body*" object.

On the consolidated basis of the electrical and magnetic observations of classical physics (which we will partly recall later in our journey), it is possible to calculate the "*spectrum*" of the "*black body*", or the intensity of radiation emitted for every wavelength by an ideal "*black body*", when the temperature is assigned. Just as an aside, let's remember that a "*spectrum*" is usually referred to as the "frequency" ν, where $\nu = 1/T$ is the inverse of the period T of the electromagnetic wave.

Some simple formulas, which, however, are non-essential, are as follows: if c is the speed of light, λ is the wavelength of radiation, the number of wavelengths in the path traveled by the light in one second is c/λ, and the number of oscillations carried out in one second is the "frequency" ν, where $\nu = 1/T$ is the inverse of the period T of the electromagnetic wave.

The paradox - which is indeed very important - is the fact that the calculated "*spectrum*" of the "*black body*" provides for an emission, at very low frequencies, of radiation with infinite energy. This cannot be, because the measurements clearly show that at some point, the actual observations differ substantially from the calculated curve. Therefore, there is some basic logical "bug". The theoretical approach must be revised *ab origine*.

Incidentally, the well-known, practically instantaneous thermometers - which measure the temperature of the human body or even of very hot objects - use the blackbody spectrum, because the maximum of the observed spectrum changes with the body temperature.

A physicist, who changes the basic rules, is usually considered by the establishment to be almost a "terrorist". Every innovator is opposed as much as possible, as he is considered a "crazy" person who subverts the dogmas and hierarchies now consolidated in the academic world, etc. In other words, it is necessary for us to find ourselves faced with a real "spoiler" who wants to antagonize, if possible, the entire academic world. This has happened numerous times since Copernicus, Giordano Bruno, Galileo, Kepler, etc.

In the case of the "*spectrum*" of the "*black body*", the "spoiler" was Max Planck (1858-1947), a German physicist, Nobel Prize winner for physics in

1918, who in 1900 introduced a "crazy" idea. That is, he enquired what happens if Nature is not continuous, or if the radiation is not represented by continuous waves, but by micro-granules (which he called "*quanta*" from the plural of the Latin term "*quantum*")?

Suddenly, the paradox vanished. Quantum Physics was born! Today, an innumerable sequence of other phenomena more than clearly demonstrates the non-continuity of natural phenomena.

In other words, replacing the data of Fig. 2.1 or Fig. 2.2 with the curve of Fig. 2.4 - as for millennia had seemed obvious, almost innate, natural, of "common sense", etc. - is the confusing "bug".

"*Continuity*" is only an illusion, a false mirage of the human mind in search of "simple" solutions - and with it also the concepts of infinitesimal and infinite. *The "continuity" hypothesis must be decisively rejected with no hesitation.*

Chapter 3

Space, Time, "Realism", ABC Theorem

"Children, let us love not in word or speech but in deed and truth. [Now] *this is how we shall know that we belong to the truth ..."*

[The first letter of John (3 18-19)]

....

The *principle of realism* maintains that the occurrence of a phenomenon is independent of the observer who observes it.

Einstein (1955; his chapter on "*The Evolution of the Concept of Natural Reality*") maintains that "*faith*" in this axiom is the basis of all natural science.

"*Any serious consideration of a physical theory must take into account the distinction between the objective reality, which is independent of any theory, and the physical concept with which the theory operates.*" [A. Einstein, B. Podolsky, and N. Rosen, 1935]. This is called the "*principle of realism*".

Nature is composed of quanta – i.e., grains of various sizes. Nature is not continuous. Conversely, geometric space and time are continuous. In fact, by definition, space and time *are not physical objects.*

Space and the memory of time are the equivalent of a continuous shelf into which we can pigeonhole our knowledge of the natural world. For example, as we will see, every living being has the perception of time in order to survive, but perhaps - unlike humans - it has a limited ability to memorize what it observes.

Space is an abstract concept of the human mind. Therefore, as a creation of the intellect, space can have properties arbitrarily selected by man - even very different from those that are generally considered intuitive. This abstract concept has the sole aim of being more or less suitable for pigeonholing the information obtained by observing Nature, just as every bookcase must be suitable for containing volumes of a given dimension, etc. There is certainly no point in listing here all the types of "space" that have been imagined by enlightened mathematicians. We simply give some simple information that is pertinent here.

Dimension of the space

The first feature of a space is the *dimension*. For instance, a linear space of dimension one (i.e., it is called a 1D space) is a simple straight line, and the "*coordinate*" of every point is the distance (positive or negative of a point P with respect to the default origin O; see Fig. 3.1).

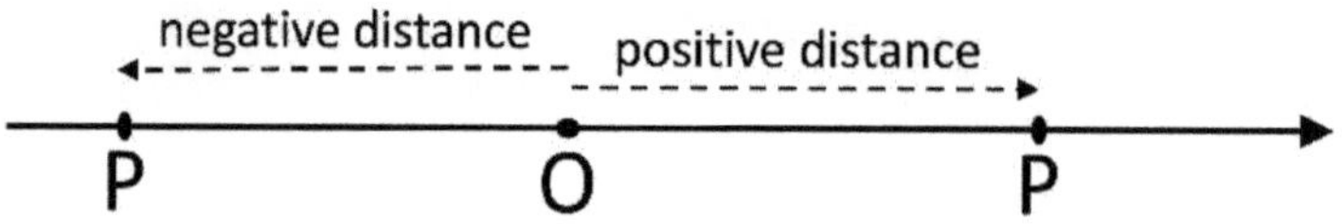

Fig. 3.1 – A linear one-dimensional space (i.e., 1D) is represented by a straight line. An origin O is defined, so every point P has its own typical coordinate expressed by its distance (positive or negative) with respect to the origin.

A space can be two dimensional (that is, 2D). Fig. 3.2a shows the best known case, represented with two *perpendicular* axes (synonymous with *orthogonal*), also called *Cartesian space*, having been used for the first time by Descartes (René Descartes, 1596-1650), a French philosopher and mathematician, among the principal founders of modern mathematics and philosophy. Every point P is described by two numbers, called "*Cartesian coordinates*".

In fact, the two Cartesian axes, also known as coordinate axes, may not be orthogonal to each other. In this case, the coordinates describing a point P can be of two types, depending on whether each coordinate is defined with lines through P parallel to the coordinate axes (Fig. 3.2b), or with lines through P perpendicular to the coordinate axes (Fig. 3.2c). Obviously, there is a broad mathematical theory (called "*tensor*" calculus) that allows you to calculate the coordinates of one type from those of the other type, but this complicated discipline is not relevant to our discussion.

It should also be considered that the coordinate axes in a 1D or 2D space in addition to being oblique (as in Fig. 3.2b and 3.2c), can also be more or less bizarrely curved. Consider, e.g., the meridians and parallels on the Earth's surface (Fig. 3.3), which is 2D, and the space thus represented is then said to be "*curved*".

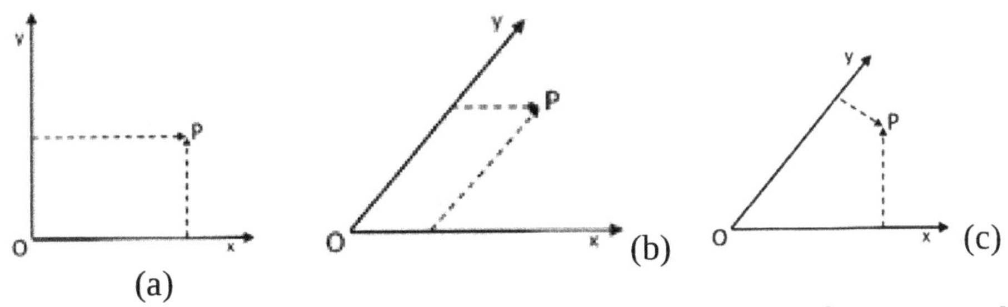

Fig. 3.2a,b,c – A linear space in two dimensions (i.e., 2D) can be represented on a Cartesian plane referred to orthogonal axes (case 3.2a), or non- orthogonal axes (cases 3.2b and 3.2c). In the case of chosen oblique axes, the relative coordinates of any given point can be defined in two ways. The concept of coordinates in a 2D flat space can also refer to variously curved coordinate axes (case not shown here). See text.

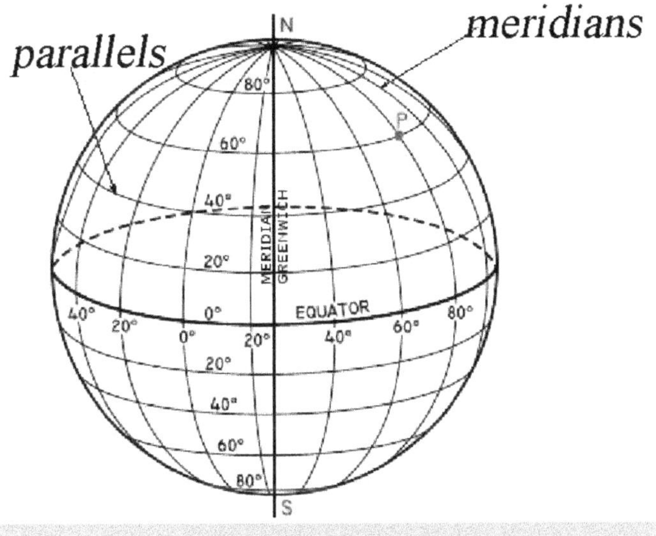

Fig. 3.3 - Simplest and best-known example of 2D curved space.

In the case of 3D space (Fig. 3.4), the same criteria apply - with orthogonal Cartesian coordinate axes as in Fig. 3.4, or even with oblique, or with variously curved axes, etc. The reader can easily imagine the large variety of coordinate axes that can be hypothesized. "*Tensor*" calculus is very powerful and allows you to manage the relationships between different definitions of 3D spaces.

All this can be generalized to the case of nD spaces with n larger than 3, obviously with no possibility of visually representing them with a figure of the type in Fig. 3.4, etc. Every space can have curvature (like the Earth's surface in the 2D case, see Fig. 3.3), or not. The "*tensor*" calculus allows you to manage everything and is part of the mathematical language. In summary, we have the possibility of imagining an appropriate space that is as suitable and as simple as possible for pigeonholing the information we have by observing Nature.

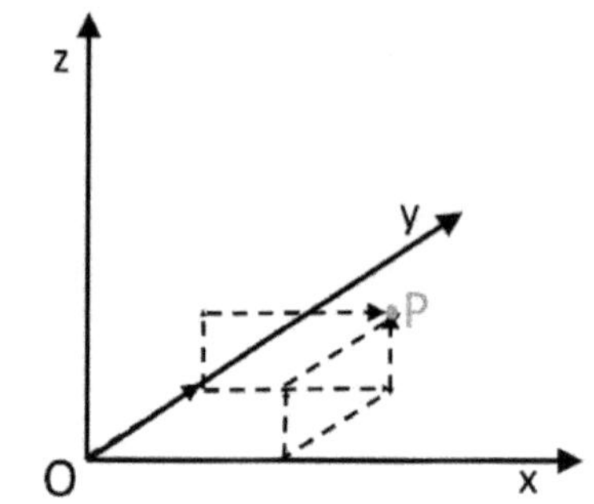

Fig. 3.4 – A linear space in three dimensions (i.e., 3D) can be represented with three Cartesian coordinate axes, in this plot hypothesized to be mutually orthogonal. Every point P is defined by three coordinates. In general, Cartesian coordinate axes could be oblique or variously curved. The concept can also be applied to hypothetical spaces with a greater number of dimensions (say, nD), even if, obviously, it is not possible to give a geometric graphic representation.

In this respect, it is worth highlighting a *very common mistake* in the scientific literature concerning general relativity. In fact, *contrary to what is often claimed* both in specialist and popular literature, 3D space in Nature is *not* curved. The possible "*curvature of space*" depends only on our arbitrary choice of whether we want to describe phenomena with a curved 3D space or not.

In other words, there is *nothing physical* in this choice, rather only a decision made by opportunity. This misunderstanding is a true logical "bug".

Space isotropy

Question: how can we *physically* and *objectively* define a straight line in the space we have chosen, while we stay in the space in which we live?

If we reflect carefully, we really have no alternative, and we must define a straight line by identifying it with a ray of light. The concept is clarified better by considering Figs 3.5 and 3.6. That is, consider a purely ideal case of a 2D world where ants live, while a hypothetical "Higher Being" knows well whether the 2D living space of the ants is Euclidean (i.e., not curved) or is curved.

Fig. 3.5 shows the case of Euclidean 2D space. The ants' perception of the flow of "time" is given by a hypothetical axis, which has notches that help the ants to detect and measure the flow of time.

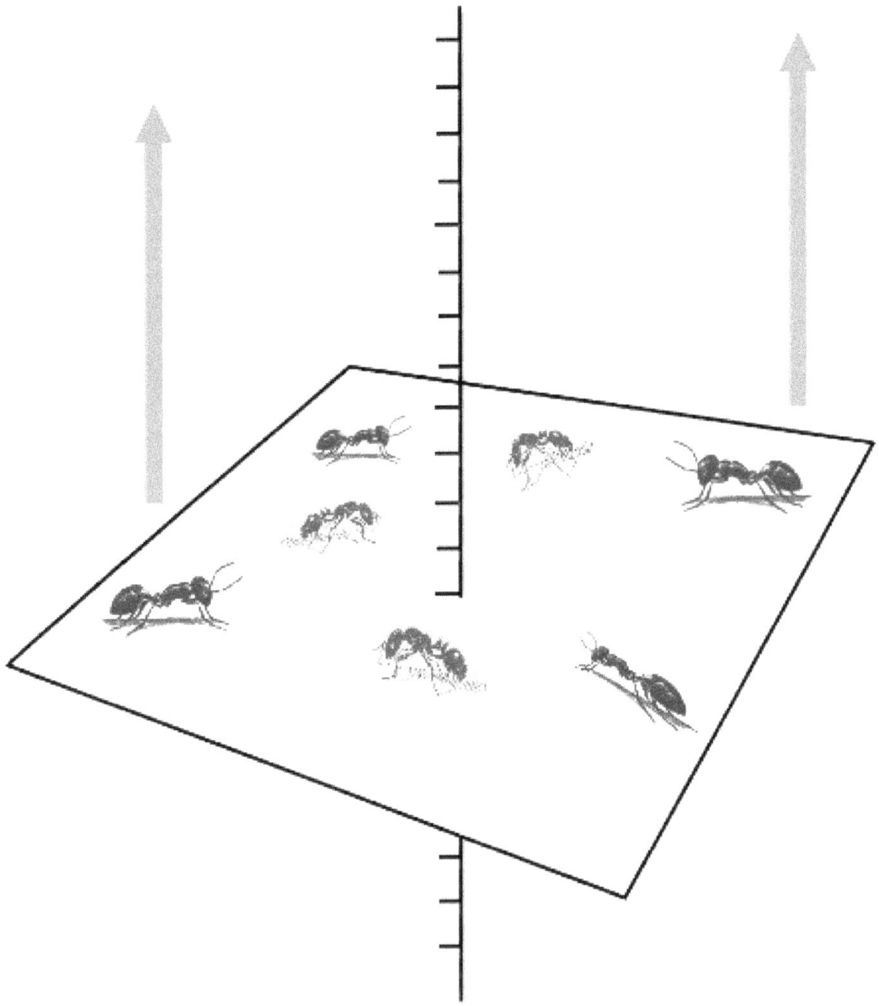

Fig. 3.5 – Ants living on a flat 2D plane can (or cannot?) judge whether their 2D living space is Euclidean or whether it is curved. The dilemma depends on the possibility of predicting some universal physical laws through which to recognize the eventual curvature of their 2D space. In this hypothetical figure their 2D space moves along a "time" axis, where some tick marks provide the ants with the perception of time flowing, while tick marks can be used by the ants to define a unit of "time". That is, these ticks are the natural clocks of the ants. See text.

Fig. 3.6 shows the case in which the 2D life space of ants is curved, although the space curvature is known only to the hypothetical "Higher Being". Even the axis that gives ants the perception of the flow of "time" is hypothetically curved, and the notches on the same axis are distributed with no apparent regularity.

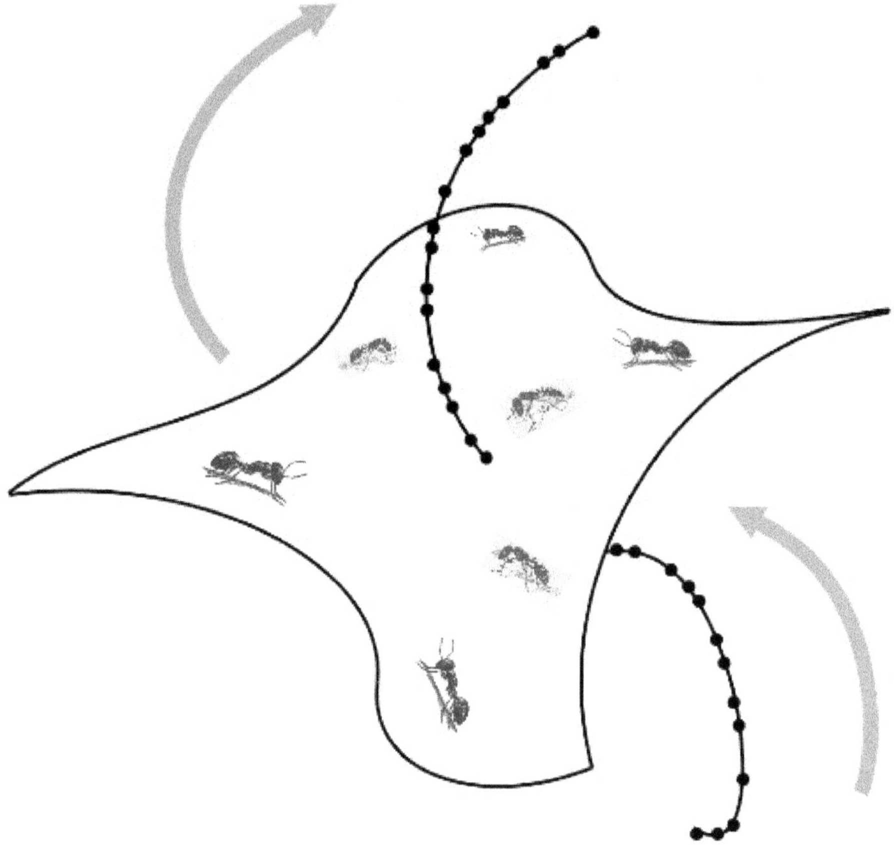

Fig. 3.6 – Ants, like in Fig. 3.5, live on a 2D surface. A “Higher Being” knows that this 2D surface where the ants live is curved, and that the axis of “time” is curved, while the signs that provide the ants’ perception of “time” are irregularly positioned along about it If ants cannot discover any universal and absolute laws, they cannot recognize that their 2D living space is curved (indeed, curved with respect to what reference system?). In fact, unlike ants, a “Higher Being” knows a “grid” of reference within an “absolute” framework that provides information on the existence of straight lines in space and time in natural reality. Only an absolute reference “grid” can be a reliable reference “clock”. In contrast, ants have no kind of alternative “grid” and “clock”. They can only believe that they live in a 2D Euclidean space, while “time” is indicated by the ticks on the “time” axis. See text.

Ants are unable to figure out whether their living space, which is 2D, is Euclidean or curved. Obviously, in the real case of our world, in the presence of gravity, the ants could have the perception of a third dimension, but this is not the case represented hypothetically by Fig. 3.5 and 3.6. If the ants use a light beam, the light propagates in a straight line in the case of Fig. 3.5, and along a curved line in the case of Fig. 3.6. In both cases, however, the ants can only assume that their space is Euclidean, with "straight lines" that are defined just by the observed paths of light rays.

Furthermore, the ants have no way of distinguishing whether the light propagates differently in the case of Fig. 3.5 or Fig. 3.6. According to the ants, their living 2D space is isotropic, i.e., light propagates identically along straight lines in all 2D dimensions - and always and only at the constant speed of light along trajectories that, by their very definition, are straight lines.

Man lives in a 3D space, but he does know about no "Superior Being" who tells him whether the space he lives in is Euclidean or curved. Man can only define a straight line in the physical world by identifying it with a ray of light, and consequently, man can only assume that he lives in an isotropic 3D space just by definition, given that due to the empirical constraint, it is strictly impossible for him to recognize if his living space is either Euclidean or not.

Man has the logical facility to speculate and define a curved space, which can be more appropriate for the purposes of particular calculations, but this is only a calculation tool, an image that may appeal to mathematicians. Furthermore, mathematics - as a creation of the mind - does not have, and cannot have, a parallel in the real physical nature of 3D space.

Realism and the ABC theorem

Let us now consider two very important concepts for our journey, the concept of "*realism*" and the "*ABC theorem*".

The so-called concept of "*realism*" claims that objective reality happens as a natural phenomenon, and is independent of whether an observer is observing the phenomenon or not.

Obviously, the observer must only carry out passive actions, i.e., he must send no signals which, in any case, can affect the occurrence and evolution of the phenomenon. Otherwise, the disturbance caused by the observer must be taken into account. The concept of "*realism*" is a "*faith*", as Einstein wrote in 1955 (the year of his passing), as recalled in the *incipit* of this chapter.

The "*ABC theorem*" is very simple and was used by Einstein - although he did not name it anything. This theorem will be fundamental for the rest of our journey. Let's look at an illustrative example.

A train is moving in one direction at the speed of $10\,m/sec$ (i.e., $36\,km/hour$). A passenger walks in the train corridor, in the opposite direction to the direction of movement of the train, with a speed relative to the train of $1\,m/sec$ (i.e., $3.6\,km/hour$).

Therefore, relative to the ground, the passenger moves with a speed of $(10 - 1)\ m/sec$ [i.e., $(36 - 3.6) = 32.4\ km/h$]. Since the passenger moves at a comparatively lower speed, in principle, he has better possibilities to observe the details of the landscape from the window.

More explicitly, we refer to three observers we will call A (Alice), B (Bob), and C (Charlie) who are in a multi-storey department store. Alice is standing on a given level of the department store, and she can distinctly read an information board. Bob, on the other hand, is on an escalator that rises rapidly at a certain speed and does not allow him to read the same board. Charlie is on the same escalator that goes up and where Bob is found, but Charlie goes down the escalator at the same climbing speed as the escalator. Charlie, therefore, always laboriously remains at the same level as Alice, while the scale rises. Charlie is therefore able to read the board with the same precision with which Alice reads it.

Based on this same principle - and in accordance with the acute analysis made by Einstein, which is a classic that we will discuss in the next chapters - if two observers move relative to each other, they could have a different perception of the laws of Nature. Conversely, if two observers are stationary relative to each other - like two passengers standing at the window in two different carriages and therefore seeing the details of the landscape identically - the natural laws that the two observers infer from their observations must be identical.

Let us now consider a general case, and let us refer to three observers: A (Alice), B (Bob), and C (Charlie). See Fig. 3.7. Every observer A, B, and C has its own reference frame with Cartesian coordinate axes. For simplicity, we consider the case of relative motions of each observer with respect to another, but always with velocity *along the same Cartesian axis* $\boldsymbol{x}$. That is, let's suppose that B moves with respect to A with velocity $\mathbf{v}$, while C moves with respect to B with velocity $-\boldsymbol{u}$, where the minus sign indicates that the motion of C with respect to B has the opposite direction to that of B with respect to A.

Therefore, C moves with respect to A with velocity $\mathbf{v} - \boldsymbol{u} = 0$, and in the case where it is $\mathbf{v} = \boldsymbol{u}$, C is immobile with respect to A, as in the case of Charlie with respect to Alice in the multi-storey department store.

In general, as it is done in the study of relativity, we consider the same phenomenon which, according to the principle of "realism", occurs independently of the observer - and which is observed in a different way by A, B, and C. But if A and C are stationary with respect to each other, the phenomena that A and C observe must be identical, as well as the natural laws that they can hypothesize to describe Nature.

That is, whatever law we hypothesize, we must *always* consider that the natural laws hypothesized by A and C must be *strictly identical*. This is the *ABC theorem*.

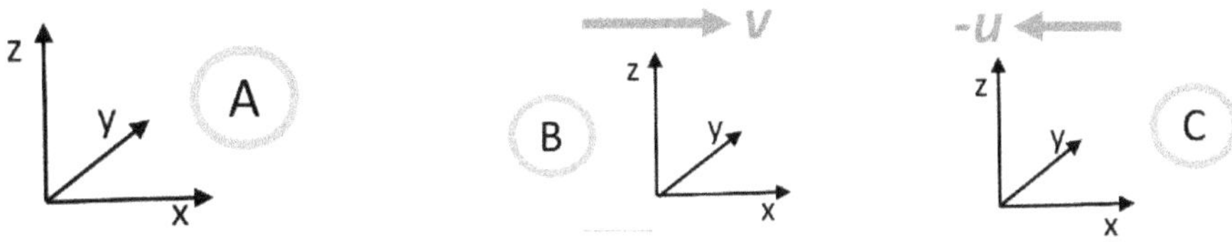

Fig. 3.7. Case of three observers, everyone symbolically represented by his own reference system which - for simplicity - is always assumed to be a 3D orthogonal Cartesian. Bob (B) moves with respect to Alice (A) with velocity $\mathbf{v}$. Charlie (C) moves relative to Bob (B) with velocity $-\boldsymbol{u}$, along the same direction as $\mathbf{v}$ but in the opposite direction, as denoted by the minus sign. Therefore, Charlie (C) moves relative to Alice (A) with velocity $\mathbf{v} - \boldsymbol{u}$. Therefore, when $\mathbf{v} = \boldsymbol{u}$, and $\mathbf{v} - \boldsymbol{u} = 0$, Alice (A) and Charlie (C) are stationary relative to each other. So the same phenomenon is observed differently by Alice (A) and by Bob (B). Conversely, Alice (A) and Charlie (C) must observe the same phenomenon identically because, in fact, they are the same observer, possibly placed in different sites. The laws of physics that Alice and Charlie deduce from observations must be identical. If a proposed law does not meet this condition, the law is not correct. This "theorem" was applied by Einstein, even though he didn't give it any name. For brevity, this argument is called the "*ABC theorem*".

Therefore, C moves with respect to A with velocity $\mathbf{v} - \boldsymbol{u}$, and in the case where it is $\mathbf{v} = \boldsymbol{u}$, C is immobile with respect to A, as in the case of Charlie with respect to Alice in the multi-storey department store.

In general, as it is done in the study of relativity, we consider the same phenomenon which, according to the principle of "realism", occurs independently of the observer - and which is observed in a different way by A, B, and C. But if A and C are stationary with respect to each other, the phenomena that A and C observe must be identical, as well as the natural laws that they can hypothesize to describe Nature.

That is, whatever law we hypothesize, we must *always* consider that the natural laws hypothesized by A and C must be *strictly identical*. This is the *ABC theorem*.

Now that we have defined the empirical constraint (Chapter 1), the false illusion of continuity (Chapter 2), and the 3D space (orthogonal Cartesian), i.e., the ideal "shelf" or "bookcase" where we can pigeonhole our observations of Nature, and that we have clarified the "physical" definition of the straight line and isotropy of 3D space, we can proceed and discuss concrete and observed phenomena.

Chapter 4

Matter and Antimatter

"Theory can be tested by experience, but there is no way from experience to the setting up of theory."

Albert Einstein (1949) (1879-1955).

....

After clarifying some basic premises - such as the empirical constraint, the illusion of continuity, and the definition of a reference space with arbitrary size and curvature - "let's get our hands dirty" with concrete observations.

"Vacuum"

The first question to ask is what matter and energy are. "*Everyone knows*" what they are (is the obvious answer). In fact, this is a response without a logical basis. "*Everyone says so*" is never the answer.

We can say that everything that has weight is "*matter*", but this presumes that we have defined what "*weight*" is.

Let us therefore begin much earlier, and consider what "emptiness" (*vacuum*) is. An obvious answer is that a "*domain*" of 3D space – i.e., a sort of given box in 3D space - is empty when it contains no "*matter*". But let's say that it cannot even contain energy, e.g., it doesn't even have to contain photons of light. That is, the "*emptiness*" (*vacuum*) is the "*nothingness*" that exists.

The definition of "*vacuum*" was widely debated and controversial among the philosophers of ancient Greece - and these were certainly not lame debates! The objection was this: if the "*vacuum*" is "*nothingness*", or something that does not exist, how can we define or describe it? Correct!

The formal definition of "*energy*" will be more fully discussed in the following chapters. For now, let's accept everybody's intuitive notions.

For now, let's consider "*matter*" and "*energy*" (we will see below how they are defined in our process of understanding Nature). Einstein wrote that "*matte*r" and "*energy*" are the same thing, and can transform into one another according to the famous formula $E = mc^2$. In reality, it almost seems as if Einstein himself found it difficult to believe such a paradoxical conclusion. Einstein would have taken the concept from a similar and equally vague idea of Olinto De Pretto (1857-1921), an assistant at the *School of Agriculture* in

Milan, and Director of the company *"Ing. Silvio De Pretto & C"*, founded in 1885 by his brother Silvio, which later became *De Pretto-Escher Wyss* in 1920. De Pretto was assassinated on March 16, 1921. Einstein believed he had "proved" this formula. In fact, he did not prove it (as we will see below).

For our logical journey, we say that $E = mc^2$ is an observed fact - and as such it is not to be demonstrated – i.e., it is a basic axiom of our knowledge of Nature, also confirmed by many known and consolidated experiments and phenomena. This was a "lightning" intuition of De Pretto and Einstein!

"*Matter*" and "*energy*" are therefore the same physical quantity, only expressed in different units of measurement (just as, e.g., I can measure a length in centimeters or inches, a distance in kilometers or miles, a surface area in square kilometers or in acres, etc.).

For simplicity, we use a new term that includes both the concepts of "*matter*" and "*energy*". We will therefore call "*emp*" – acronym of "*energy, matter, primordial*" – something that exists and which can be both matter and energy, expressed in predefined measurement units.

We will then say that something "*exists*" if it has a non-null content of "*emp*". Otherwise, when there is no "*emp*", we have "*nothingness*", the "*non-existent*", or rather the "*emptiness*" (*vacuum*).

In this respect, a "*physical principle*" - which (at least for now) cannot be fully proven but which seems to be universally respected - is the "*principle of the maximum possible emp density*". That is, whenever some natural process leads Nature to concentrate emp in a given domain of space and time, with a local density higher than a maximum threshold (which can be even unknown, but it exists), some phenomenon must necessarily happen such that the system can expel from that domain the amount of emp in excess of the given threshold.

It seems like this principle always holds true. We note that the only exception could be the so-called "*black hole*" which would swallow up everything by gravitational attraction. In fact, the concept of "*black hole*" is highly controversial - even among most authoritative scholars - and in any case, it is an extreme case resulting from speculative (mathematical) arguments.

A non-secondary and important aspect connected to the "*principle of the maximum possible density of emp*" is the existence of an "*absolute reference system*", i.e., the *æther*. In fact, kinetic energy (formally defined later in our journey) must be considered among other contents of emp. If æther does not exist, then changing the reference system would be enough to violate the "*principle of the maximum possible density of emp*". But this is contrary to "*realism*" (see Chapter 3). In this journey of ours, it will therefore be assumed that æther exists, as we will see in detail later.

"Matter" and "energy"

Let us therefore see how the concepts of "*matter*" and "*energy*" have entered our common way of thinking. A historical excursion is necessary, starting from the science according to Aristotle, also taken up entirely by Dante Alighieri (1265-1321) – and *La Divina Commedia* (*The Divine Comedy*) was for centuries the equivalent of the *Treccani Encyclopedia* for the entire cultured world of Europe and for the entire world of the time.

For Aristotle, the Earth is at the center of the Universe. The four elements (earth, water, air, fire) are by their intrinsic property ("*virtue*") arranged in this order from bottom to top - and above all, the celestial spheres are located, which contain various stars, up to the last sphere, that of the *Empyrean*, which is the seat of God.

Pope Leo III, on Christmas night in 800 AD, gave the "*divine*" investiture to Charlemagne (742–814), king of the Franks from 768, king of the Lombards from 774, and from 800 the first Emperor of the Romans, crowned in the ancient Basilica of St. Peter in the Vatican. The aim was to bring some order to those very tormented times.

This, however, was later understood as a "*divine*" investiture of the Emperor and the related nobility. Since then, anyone who had questioned the centrality of the Earth in the Universe was implicitly questioning this investiture. This man was therefore comparable to a heretic, a sort of terrorist etc.

For Aristotle, an object moves due to the intrinsic, metaphysical property of that object, i.e., due to a sort of "*virtue*" that is congenital, innate, metaphysically justified. The radically opposite concept was that of "*inertia*", by which a body moves only if some cause alters its motion. If this cause does not exist, the body remains in its state either of rest or of uniform motion.

Galileo had perfectly understood the correctness of the principle of inertia - in sharp contrast to Aristotle - and this had probably also been well understood by Leonardo da Vinci two centuries before Galileo. However, Leonardo and Galileo missed a complete and self-consistent logical structure - i.e., without internal contradictions - with which to construct the entire logical edifice of the science of Nature. A complete, self-consistent logical settlement was achieved only by Isaac Newton (1642-1726), an English mathematician, physicist, astronomer, natural philosopher, theologian, historian and alchemist.

Newton's three principles of dynamics were a "lightning" idea. The first principle was that of inertia, and the third principle was that of action and reaction (i.e., if one object causes an action on another object, the second object causes an equal and opposite reaction on the first object). The second principle of Newton's dynamics - perhaps the most problematic - needs to be explained.

Newton began with the definition of "*force*". He considered a spring, or a metal thread, ideally elastic. The spring or thread stretches more or less, depending on the "*force*" applied. So he had an empirical way of measuring a "*force*". Then he considered any body, either solid or otherwise, and with his second law, Newton established that the force $\boldsymbol{F}$ and the acceleration $\boldsymbol{a}$ (i.e., the variation in speed undergone by the body) are related by the relationship $\boldsymbol{F} = m\boldsymbol{a}$. In other words, the constant m (which Newton called "*mass*") is simply an index of a sort of "laziness" of the body in responding to the stress due to the force $\boldsymbol{F}$. The more "lazy" the body is, the less it responds. That is, this definition of "*mass*" is connected to the "*inertia*" that is associated with the body. For this reason, this definition of "*mass*" is called "*inertial mass*".

An unsolved problem – and one that was very relevant in those years – concerned the motion of the planets. Tycho Brahe (1546-1601), Danish astronomer and astrologer and supporter of the geocentric hypothesis of Ptolemy and Aristotle, had given his extensive collections of astronomical observations - made from the *Uraniborg Observatory* - to his brilliant assistant Johannes Kepler (1571-1630), German astronomer, mathematician, astrologer, natural philosopher, and writer of music. Johannes Kepler - a supporter of the heliocentric hypothesis with Galileo and Copernicus - did a truly gigantic, incredible job, while living in Prague, carrying out an empirical analysis of the data, and thus managed to formulate his famous laws of planetary motion. For the rest of his life, Kepler desperately - and in vain - searched for an explanation for his empirical laws, relying (if possible) on magnetic interaction.

Thus, *57* years after Kepler's death, Newton "invented" "*universal gravitation*". That is, any two celestial objects attract one another with a force proportional to their respective mass M_1 ed M_2, and inversely proportional to the square of their distance r. In formulas $\boldsymbol{F} = G\,M_1 \times M_2/r^2$, where G is an appropriate constant, called gravitational constant. This explanation is still considered topical, although with some important logical shadows - and associated paradoxes, as we will see below.

It must be emphasized that the mass that appears in the formula of gravitational attraction is considered identical to the "*inertial mass*" defined above in the second law of dynamics. But in principle this new definition of mass could also be different – and this new definition is therefore called "*gravitational mass*".

We will return to these very important issues later. For the discussion of this chapter, it is enough for us to have clarified the meaning of "*mass*", of "*matter*" as something "*that exists*", and something "*that exists*" can be considered the definition of "*matter*". Everything else concerns mathematical "technicalities" consequent to Newton's axiomatic approach as described above.

Newton's three principles of dynamics had an unexpected impact - and *are the backbone of all the development of science and engineering until the discovery of quantum physics* (in the first decades of the 1900s). Newton was in fact, the true first creator of the colossal revolution that occurred during and following the Enlightenment. The new progress in science and engineering required, especially during the 19th century, the development of the wonderful logical "monument" of the mathematics of continuous functions (the so-called differential and integral calculus). The quality of life for everyone underwent a radical change. The great absolute monarchies became unsuitable for the new world dominated by the science of Nature, and by a little more than a century they fell - if not becoming constitutional monarchies. Newton caused a true cultural revolution and more!

Technicalities

For the more curious reader, we give a brief reminder of some of the technicalities mentioned above. One technicality concerns the definition of potential energy and kinetic energy - even if, for now, all this is not strictly indispensable for our journey.

Let us look at a concrete example. Let's consider a high-altitude reservoir of a hydroelectric plant. A quantity m of water is introduced every second into the penstock of the hydroelectric plant. The water falls into the pipe until it reaches the speed $\mathbf{v}$. Water initially has a greater "potential energy" the higher the starting altitude. On the contrary, at the end of the penstock, the water has lost its "potential energy" - which it had when it was at high altitude - and which has now been transformed into "kinetic energy". With Newton's principles of dynamics, it is shown that the "kinetic energy" is $(1/2)m\mathbf{v}^2$. Therefore, the water impacts the blades of a turbine, which drives a "dynamo" and produces electricity, which is fed into the distribution circuit for the users. That is, the original "potential energy" has been transformed into "kinetic energy", and finally, the "kinetic energy" has been transformed into "electrical energy" which is consumed by users as electrical energy.

Let's take another concrete example: two colliding billiard balls. The balls have mass m_1 ed m_2, respectively, and initial velocities $\mathbf{v}_1$ and $\mathbf{v}_2$ - i.e., the second ball can also be in motion, or not (relative to the billiard table). Note that the speeds in the formulas are in bold to indicate that they also contain the direction of movement of the ball. Let's call $\mathbf{v}_1'$ and $\mathbf{v}_2'$ the speeds of the two balls after the collision. With the principles of Newton's dynamics it is demonstrated that "kinetic energy" must be conserved, i.e.

$$\left(\frac{1}{2}\right) m_1\mathbf{v}_1^2 + \left(\frac{1}{2}\right) m_2\mathbf{v}_2^2 = \left(\frac{1}{2}\right) m_1\mathbf{v}_1'^2 + \left(\frac{1}{2}\right) m_2\mathbf{v}_2'^2 \qquad (4.1)$$

Furthermore, it is shown that momentum must be conserved, where "*momentum*" is defined by the product of the mass times the velocity of every object. That is, it must always be

$$m_1\mathbf{v}_1 + m_2\mathbf{v}_2 = m_1\mathbf{v}_1' + m_2\mathbf{v}_2' \qquad (4.2)$$

where, in fact, the relation (4.2) is composed of two relations between the two components of the balls motion on the billiard table.

It is shown that the relations (4.1) and (4.2) - which are called, respectively, *conservation of the total energy* of the system, and *conservation of the total momentum* of the system - must apply, according to Newton's principles, to any collision between any two or more objects.

For the sake of completeness, we also remember that in the case of balls to which a rotational motion is imparted - as the most skilled billiard players do - it is also necessary to consider *the conservation of the "total angular momentum"* of the system, similarly defined, in order to quantitatively describe the rotation of the balls. We will not specify this concept in detail here to avoid making our text heavier.

In any case, these "technicalities" apply to the formulation based on the three principles of Newton's dynamics which - later - we will consider only as "statistical" aspects of a phenomenon. The phenomenon, on the contrary, should be described regardless of continuous functions, as fully specified below.

Antimatter

What now remains to be defined is "antimatter".

It is a simple experimental fact, an observed and incontrovertible matter-of-fact. The first observation of a "positive electron" called "*positron*" - which today must be more appropriately called "*antielectron*" - dates back to 1932, made by Carl David Anderson (1905-1991), an American physicist, Nobel Prize winner for physics in 1936. Antimatter had been hypothesized in various forms by authoritative theoretical physicists. Perhaps the best known was Paul Adrien Maurice Dirac (1902-1984), a British physicist, Nobel Prize winner for physics in 1933 (together with Erwin Schrödinger) for "*the discovery of new fruitful forms of atomic theory*", who made fundamental contributions, formulating, among other things, the homonymous equation and predicting the existence of antimatter. Dirac is considered one of the most important physicists of the 20th century. Today, particle/antiparticle pairs (not just electron/antielectron pairs) have been observed experimentally by numerous highly authoritative scholars. They are now a well-established custom. Let's not go into detail here.

The common principle is shown in the explanatory scheme in Fig. 4.1.

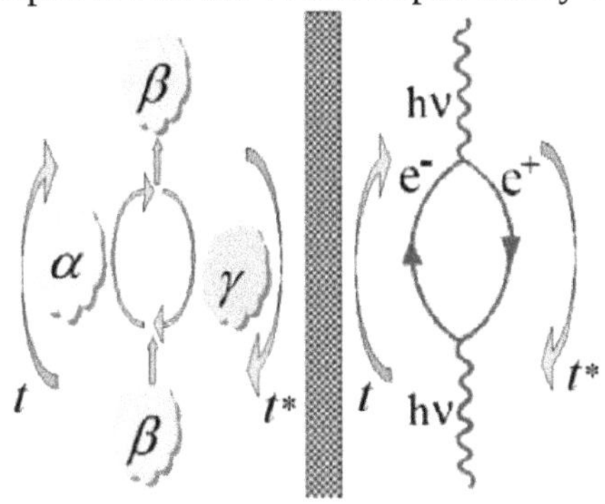

Fig. 4.1. In the right panel, a photon with energy $h\nu$ is indicated, which materializes in an electron (e^-) and positron e^+ (more properly called antielectron) pair. The phenomenon is paradoxical and unexplained, and is observed regularly, but the phenomenon also concerns other particles, other than electrons and antielectrons. This is a mystery that we need to explain. All of this must be interpreted (left panel) with the existence of three worlds. The "absolute" world is the β world that is timeless. Occasionally, from the world β, objects are generated in two symmetrical and mirrored worlds α and γ. In world α time passes as we know it. In the γ world, time flows in the opposite direction, from the future to the past, i.e., we have an antitime. This logical scheme seems to be the only possible one to solve the mystery of antimatter. The concept of a timeless "absolute" world is not new, but it is found in *Genesis* and in sacred Hindu texts (see text and Chapter 9).

Some theoretical hypotheses more or less timidly suggested the existence of another world - symmetrical to ours - where time perhaps flows from the future into the past, with perhaps negative masses, etc., a true paradox. In 2005, Giovanni P. Gregori hypothesized that "time" is not absolute - as we have taken for granted since the days of the ancient Greek philosophers. The "absolute" world is rather a world without time. Occasionally, symmetrical objects are generated from the timeless world, one in our world with time as we know it, and one in a symmetrical world where time flows in the opposite direction (i.e., it is antitime) and where masses and energies are negative. That is, a perfect dichotomy exists in Nature.

In Fig. 4.1, our world with time is called world α, while world β is the world without time, and world γ is the world symmetrical with antitime.

Note that in a world with no time, it is not possible to number objects - since there is no before and after. It is not possible to define a speed (which is a displacement divided by a "time", but there is no "time" or "antitime"). So photons - which are part of the absolute β world, which is without "time" and without "antitime" - have no speed in the world β. Photons, however, are observed in world α moving with the constant speed c of light. Symmetrically, the photons observed from the γ world move with the speed of light (but referred to antitime).

By delving into the archæology of the concept of time, we find that the reference to an absolute world without "time" is explicit in *Genesis*, and also in Hindu sacred texts. The debate between Christian thinkers advocating an absolute world without "time" - as opposed to Greek thinkers advocating a world with absolute "time" - is still found in the 3rd century AD (especially with Origen). Today, this view survives in the world of the mystics.

It is paradoxical that the ancient concept of an absolute world without "time" – as opposed to classical Greek thought, which held the world as it is now generally regarded as having absolute "time" – has now been rediscovered to get rid of the paradox of physics associated with the existence of antimatter.

In conclusion, we must consider the existence of *antimatter to be an incontrovertible fact*. It must be accepted, and the only way to explain it is with the hypothesis symbolically expressed by Fig. 4.1.

Over the course of our journey, we will see how matter and antimatter are distributed in the Universe. But for now, let's limit ourselves to these experimental observations repeatedly performed in laboratories.

A photon of sufficiently high energy occasionally "*materializes*" into a particle and antiparticle pair. Each antiparticle subsequently - very quickly - "*annihilates*" with a particle of the symmetrical specular world, thus returning to being a photon. The experimental fact is incontrovertible – observed as a customary rule, e.g., with the so-called "Wilson or cloud chamber" [Charles Thomson Rees Wilson (1869-1959), British physicist and inventor, of the cloud chamber in 1911, who shared with Arthur Holly Compton the 1927 Nobel Prize for Physics], or with the "bubble chamber".

But the explanation for "*matter*" and "*antimatter*" remains a mystery. Our journey will explain this too.

Anticipating the conclusion, note that we live in the world α where time flows in one direction only, and disorder increases (in technical terms, it is said that *entropy* always increases, and this represents a further classic paradox to be resolved). Only occasionally do we manage - with very particular experiments - to generate antimatter for very brief instants, which almost immediately "annihilates" upon returning to the β world.

But there would be entire galaxies or clusters of galaxies made up of matter alone. However, symmetrically, there would be galaxies or clusters of galaxies made up of antimatter alone. The first hypothesis of galaxies and antigalaxies was made by Hannes Olof Gösta Alfvén (1908–1995), a Swedish physicist, Nobel Prize winner for Physics in 1970, together with the French physicist Louis Eugène Félix Néel (1904-2000). Alfvén formulated (in 1950) the theory of magneto-hydro-dynamics (*MHD*), or the dynamics of fluids (liquids and gases) when they are electrically charged - and today it seems incredible that previously for centuries official science had only dealt with electrically neutral fluids.

One day it will happen that galaxies or clusters of galaxies made of only matter will "*annihilate*" with galaxies or clusters of galaxies made of only antimatter, and both will enter the β world. Subsequently, from the β world new galaxies and clusters of galaxies of only matter together with others of only antimatter will re-emerge due to "*materialization*", and the cycle will continue.

Note that the Christian religion foresees the end of the world and the "resurrection of the flesh" as an act of Faith, while the Hindu religion foresees "reincarnation", but this too will be clearer at the end of our journey.

Having clarified the meaning of "*matter*" and "*antimatter*", we enter more and more into the heart of the debate.

Chapter 5

Electromagnetism and Three "Original Sins"

"I do not know what I may appear to the world, but to myself I seem only to have been like a boy, playing on the seashore, and diverting myself in now and then finding a smoother pebble or a prettier shell than ordinary, whilst the great ocean of truth lay all undiscovered before me."

[1727, Sir Isaac Newton (1642-1727)]

....

Three original sins

"*Electromagnetism*" is a word that scares many, who consider it a concept that is difficult to understand. Let's try to dispel this "myth".

In the previous chapter, we saw how Newton definitively brought "common" thinking out of the misunderstandings of Aristotelian science, inaugurating an unprecedented era of applicative developments which did not encounter logical difficulties until the "discovery" of quantum phenomena by Max Planck in 1900 (see Chapter 2).

Newton's formulation, with all its merits, however - unconsciously - had three intrinsic "bugs" which later turned out to be very important, three true "original sins":

1) Newton treated natural phenomena - which are essentially non-continuous, i.e., quantized - with *continuous* functions. That is, he replaced observations of the type in Figs 2.1 and 2.2 with data of the type in Fig. 2.4. Let's say that Newton applied a "statistical" interpolation to intrinsically "discrete" phenomena ("discrete" means everything that is divided into non-infinitesimal quantities, i.e., that is not "continuous").

2) Newton introduced the concept of "*force*" with reference to a spring or a thread, in a very practical, approximate way, although very effective for dealing with mechanical phenomena, but without too many worries about the logical foundations of knowledge. This was a way that worked anyway, and that was enough for him.

3) Newton, when introducing universal gravitation, imagined that the gravitational signal propagates *instantly* from one body to another (we will return to this crucial point later).

The success of Newton's formulation - a true "sacred monster" of human history - was and still is such that no one has ever dared to seriously question it. This triple original sin influenced the subsequent formulation of electromagnetism, which developed in the image and resemblance of Newtonian mechanics.

However, with subsequent development, electromagnetism ended up first highlighting two of these "bugs" - namely the still unsolved paradox of the wave/photon dilemma (linked to the *continuity* hypothesis), and the electromagnetic signal which does not propagate instantaneously from one body to another, but has a finite speed (the speed of light).

However, the concept of "*force*" has still remained unchanged in use, even though theoretical physics is increasingly looking at "*Feynman graphs*" [Richard P. Feynman (1918-1988), American physicist and scientific communicator, Nobel Prize winner for physics in 1965 for developing quantum electrodynamics]. With quantum electrodynamics, Feynman rejected the concept of waves - and all phenomena are explained with a sort of tiny "bullets", or "quanta", exchanged between different particles - particles that are themselves quanta. However, in fact, the supporting structure of quantum electrodynamics - the frontier of today's most advanced theoretical physics - always inexplicably begins with the use of continuous functions, and then quantum electrodynamics partially "converts" to the use of discrete quantities (the "Feynman graphs"). But this was not enough - and therefore everything is still the subject of important studies and debates.

A final solution still seems far off. On our journey, we will explain all these still unsolved paradoxes - but let's go step by step and put our ideas in order.

Some pivotal experiments in the history of physics

Let us begin by briefly recalling some "historical" experiments - observational matters-of-fact - which determined the birth and development of electromagnetism up to its formal definition, which would now seem logically self-consistent and impeccable – i.e., an "axiomatic" formulation, that is based on axioms formally correct according to the acute definitions of James Clerk Maxwell (1831-1879), Scottish physicist and mathematician.

The "historical" experiments are: *(1)* rubbing electricity, *(2)* the Voltaic pile, *(3)* the magnetic effect of an electric current, and *(4)* electromagnetic induction. Maxwell and Poynting then followed, and the whole matter was thus clarified - but new important problems arose. But let's proceed in order.

Rubbing electricity (triboelectricity)

The first knowledge of electrical phenomena dates back to times preceding ancient Greece. By rubbing two objects together, you can see that the two objects attract each other. The most typical case is rubbing an amber object with a piece of woolen fabric or other.

The passage of something which was later called electric charge was hypothesized. The first studies probably date back to Thales who studied in 600 B.C. the electrical properties of amber called elektron (ἤλεκτρον), and from this term the word "electricity" derives. Plato describes electricity in Timæus (written in 360 BC).

The phenomenon is common to all materials, not necessarily amber and cloth. The definition of positive and negative electrical charges was the result of an arbitrary, conventional choice.

Today, everyone knows that matter is made up of molecules, and every molecule is made up of atoms. Every atom is like a sort of miniature planetary system, with electrons (with a negative charge) orbiting around an atomic nucleus (with a positive charge), and paradoxically and inexplicably, the electrons do not precipitate onto the nucleus. We will explain this below too.

Frictional or rubbing electricity is called "*triboelectricity*". The phenomenon can be explained by the fact that rubbing slightly heats the surfaces of the two materials. Some electrons are freed from the original links within the atom or molecule, and - once free from the previous links - they are attracted to the medium that has the most affinity with them, or rather they choose the medium with the greatest "*electroaffinity*".

Experiments have thus been carried out (since the time of Thales), and it has thus been experimentally seen that charges of the same sign repel each other and charges of different signs attract each other ["Coulomb's law", by Charles Augustin de Coulomb (1736-1806), French physicist and engineer]. The law is similar to that of gravitation, replacing masses with electric charges and also replacing the constant G with a new suitable constant.

The first knowledge of magnetic phenomena is due to rocks (magnetite) that are oriented towards the magnetic north pole. The phenomenon was certainly also known in the West, but the first applications - as a "compass" - appear to come from ancient China, and was imported into the West. But let's not get into these details. Bars of magnetite were the first sources of magnetic "force" available to experimenters.

Alessandro Volta and the Voltaic pile

Triboelectricity was practically of little use. To understand the significance of the following discovery, consider a relevant example.

Let's suppose we drop minute grains of sand into the atmosphere from a height of $300 - 400\ m$. In practice, they would be of no use, apart from at most a negligible nuisance. Suppose instead that a boulder falls from $10\ cm$ in height. If we use the impact of the boulder we can get something practical out of it, especially if we use several boulders together and if we can make them fall from higher heights. This is what Alessandro Volta (1745-1827) did in 1799 - a chemist and physicist born in Como, inventor of the Voltaic pile, and discoverer of methane gas.

For simplicity, we only describe the first historical pile (which today is called "Voltaic"; Fig. 5.1), and it is made up as follows:

- a vertical wooden stand on a circular base
- copper and zinc discs
- a cloth soaked in a mildly acidic solution, made up of water and sulfuric acid, and placed between copper and zinc disks
- two copper threads.

The copper and zinc discs are placed alternately, in the order copper-zinc-wet-copper-zinc, and so on, all kept vertically by the wooden structure. Once the disks and the damp cloth have been placed on the support, by connecting the first and last disks of the column with two copper threads, a passage of current is created between them.

Fig. 5.1 – A pile by Alessandro Volta now in the *Tempio Voltiano* in Como.

Volta first believed that the flow of current was due to the simple contact of the two metals. Then he understood that the passage of current was caused by the chemical reactions caused by the damp cloth. In fact, during the functioning of the battery, the zinc was consumed while the copper remained intact or possibly oxidized. Zinc therefore gives up two electrons, and these electrons - contrary to what one might think - do not pass to copper, but to the H_2O^+ ion which was formed by the dissociation of sulfuric acid in water which removed an electron from the water molecule (H_2O).

The resulting device allowed Volta to produce an electric current – which he measured by the contraction of the muscles of a dead frog. Today batteries are based on other, higher-performance materials, but the principle is the same, based on a chemical process.

The conventional symbol of a pile or battery - whether Voltaic or not, or more generally of a battery of piles or batteries - is shown in Fig. 5.2a. In the above example of the grain of sand and the boulder, the triboelectric phenomena are the equivalent of useless grains of sand, while the pile is the equivalent of the boulder. In general, it is possible to place numerous piles "*in series*" as in Fig. 5.2b, or "*in parallel*" as in Fig. 5.2c. Placing many stacks in series is like dropping a boulder from a height many times greater than a single stack. Placing many stacks in parallel is like dropping many boulders simultaneously from the same height.

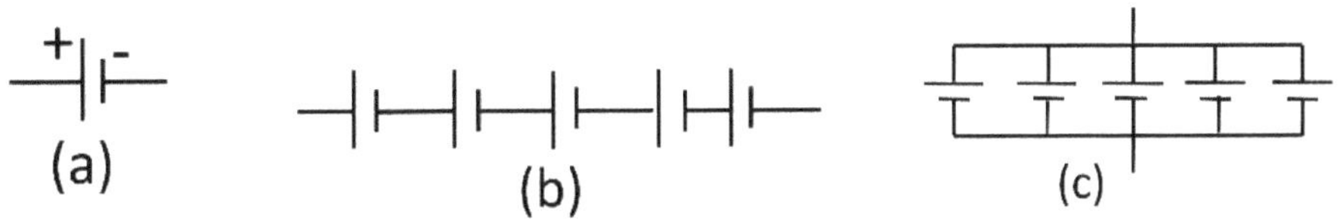

Fig. 5.2 – *(a)* Conventional symbol to indicate a Voltaic battery, or more generally a battery of cells anyway. *(b)* Example of 5 stacks in series. *(c)* Example of 5 stacks in parallel. See text.

A unit of measurement was also defined to express the height from which the boulder falls - in more correct language, it is called the "*potential*" of the pile. The chosen measurement unit is called *Volt* in honor of its discoverer.

The Voltaic pile was one of the most important discoveries of all, a "lightning" discovery, for the study of electromagnetism. Indeed, one can actually say that electromagnetism could only be born after this discovery. By placing several piles in series or parallel, it was possible to generate a large electric current. Finally, it was possible to have a significant flow of electrons and carry out experiments that were previously unthinkable.

André-Marie Ampère and the magnetic effect of electrical current

Consider Fig. 5.3. An electric current flows within a conductive cable. Don't forget that in reality it is always the negatively charged electrons that move from the negative pole to the positive pole - in order to compensate and make the difference between the two poles neutral. It is said that the conductor cable was thus short-circuited between the two poles. If the power generator is sufficient – i.e., the number of piles is large enough – the electric current can be sustained for a long time.

By exploring space with the aid of a magnetized bar (such as the needle of a magnet), the French physicist André-Marie Ampère (1775-1836) discovered that an electric current generates a magnetic signal aligned along concentric circles with the axis of the electric current. This discovery had enormous consequences. Ampère is now honored by naming the unit of measurement of electrical current after him.

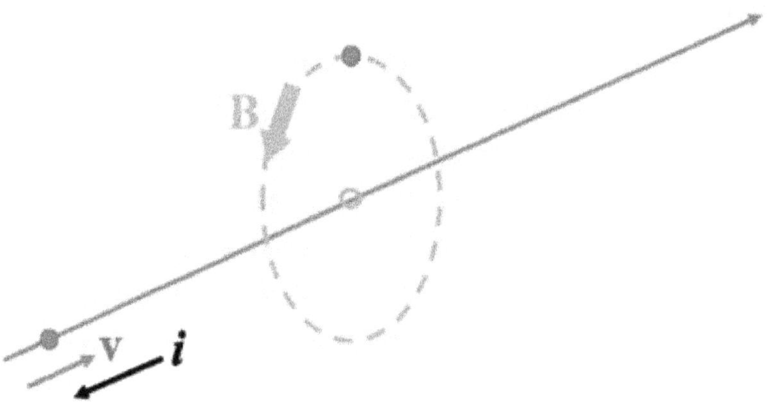

Fig. 5.3. A fixed electron, or even the needle of a magnet, represented by a small ball at the top of the figure, measures a magnetic "force" $\boldsymbol{B}$ (large arrow) generated by an electric current $\boldsymbol{i}$ shown by the black arrow. The electric current is the same as a flow of numerous electrons with velocity $\mathbf{v}$ (arrow opposite to $\boldsymbol{i}$) flowing inside an electrically conductive cable. The direction of $\boldsymbol{B}$ reverses when you reverse the direction of the electric current. The intensity of $\boldsymbol{B}$ decreases with $1/r$ where r is the distance of the fixed electron from the current-carrying cable.

Ampère discovered that the intensity of the magnetic "force" generated by the current grows in proportion to the intensity of the current and decreases as the radial distance r from the current in proportion to $1/r$.

The discovery meant that a new source of magnetic "force" was available – i.e., a current-carrying circuit rather than a magnetite bar - to be used in experiments.

Michael Faraday and electromagnetic induction

Let us consider the magnetic "force" generated by one or more sources, and imagine the same magnetic "force" as a flow, almost as if it were a flow of water. Suppose we intercept this flow of magnetic "force" with an electrical conductive circuit, e.g., with a metal ring that is moved through space. The British physicist, chemist, and popularizer Michael Faraday (1791-1867) thus discovered that - by varying for any reason the flow linked by the metal ring - the circuit of the metal ring is crossed by a current that tries to oppose the change in magnetic "force" flow. This is the phenomenon called *electromagnetic induction.* We will return to this concept in greater detail in Chapter 14.

If we *compel* a metal circuit or ring to move within a space where there is a magnetic "force" varying from point to point, we have created a generator of electric current - and this is called a "*dynamo*".

If instead we let the circuit *move freely* to try to minimize the "disturbance" caused by the flow of magnetic "force", we have created an "*electric motor*".

Obviously, it took many decades to build practically usable dynamos and electric motors, but this is not of concern for our logical journey.

The principle is illustrated symbolically in Fig. 5.4.

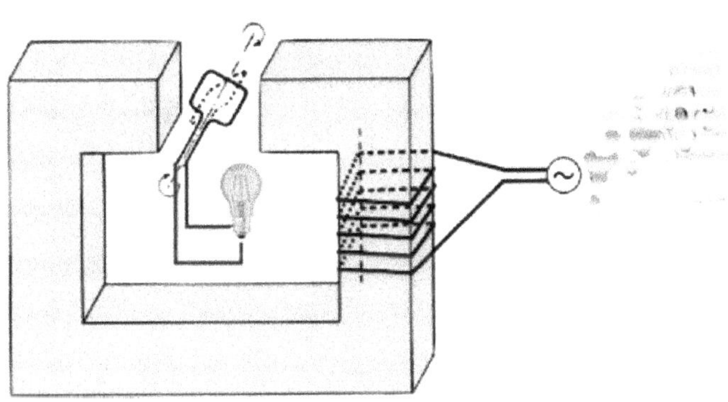

Fig. 5.4. General scheme of a dynamo. An enormous ring (shown as a square section) of iron, called a stator, is magnetized by a current shown here by the coil on the right. The stator has an interruption within which a moving circuit is found. The magnetization of the stator must take place with direct electric current. Moreover, it is possible to imagine a direct current that can vary (e.g., due to electromagnetic induction by the Sun shown at the right), thus modulating the magnetization of the stator, hence the efficiency of the dynamo. The moving loop is rotated between the two stator expansions, e.g., by the water that drives the blades of a turbine of a hydroelectric power station. The electric current generated through electromagnetic induction in the mobile circuit is distributed to the user network.

James Clerk Maxwell and the concept of "field"

Experimental knowledge was a challenge, and an axiomatic logical framework was needed. James Clerk Maxwell (1831-1879), a Scottish physicist and mathematician, succeeded in doing so, but he made use of a very bold new concept which - in many respects - repudiated the idea of "action at a distance" used by Newton, i.e., Maxwell was a "spoiler" for the science of the time.

Newton hypothesized an instantaneous (gravitational) interaction, but the same concept was implicitly adopted also for electrical and magnetic phenomena. Maxwell instead hypothesized that a source of electric or magnetic "force" *modifies locally at every point of all space* up to an "infinite" distance - in the sense that, *when a sensor capable of detecting the presence of the "force" is placed* in a point, then the sensor detects the "modification" undergone by the space at that point.

Maxwell called the space thus modified a "*field*". Thanks to this acute hypothesis, Maxwell was able to write his famous "*Maxwell equations*" which give a very effective synthesis of all known experiments, even though within the presumption of continuity. It was a "lightning" discovery, of enormous importance.

Since then, we no longer speak of electric or magnetic "force", we always speak of "*electromagnetic field*". Among other consequences, we spoke of the density of electric and magnetic energy, which is deposited in "infinite" space. That is, we no longer worry about the sources of the electric or magnetic "field". What matters is the value of the "field" stored in a point. The "field" is continuous, and Maxwell was thus able to write equations applicable to every "infinitesimal" micro-volume of space, without having to worry about how the "field" was generated.

Electromagnetic Radiation (Poynting Vector)

By solving Maxwell's equations in the simplest case, it was discovered that space is crossed by wave signals, real waves, of the form schematized in Fig. 5.5.

The propagation of an electromagnetic wave can be effectively described with a vector, called "*Poynting vector*", by the British physicist John Henry Poynting (1852-1914), who, in 1884, was among its discoverers. The "*Poynting vector*" expresses the straight line propagation of the signal in the shape of a wave, and expresses precisely the flow of energy that propagates in space. We will come back to "straight line" propagation later.

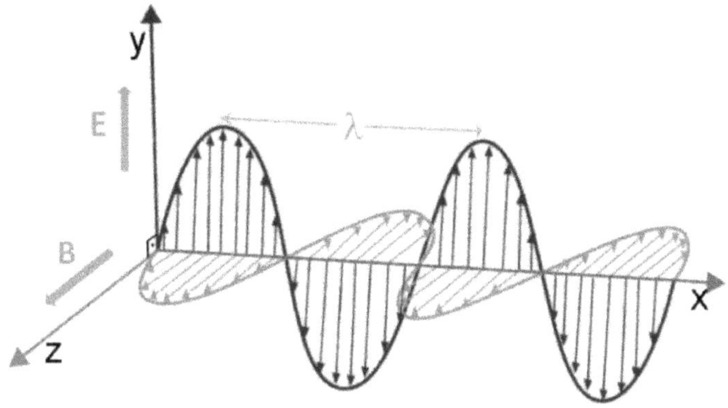

Fig. 5.5 – Calculated electromagnetic wave, propagating to the right, with electric field $\boldsymbol{E}$ along the $\boldsymbol{y}$ axis and magnetic field $\boldsymbol{B}$ along the $\boldsymbol{z}$ axis. The associated Poynting vector points in the $\boldsymbol{x}$-axis direction. Notice that when $\boldsymbol{E}$ is directed upwards in the figure, then $\boldsymbol{B}$ is directed towards positive $\boldsymbol{z}$. See text.

The signal has a variable wavelength λ depending on how it is generated - but the velocity is always the same. Great was the amazement when it was discovered that this speed – mathematically calculated on the basis of laboratory experiments on electricity and magnetism – was exactly identical to the speed c of the propagation of light measured directly with a lighthouse!

It was then deduced that light is an electromagnetic wave of very short wavelength. The waves of longer wavelengths are those used for radio and television communications, etc.

However, a critical and crucial aspect concerns the reference system. Electromagnetic waves (and light) have velocity c, but with respect to which reference system? Does changing the reference system change the speed of light? In other words, seeing Fig. 3.4, we ask ourselves whether the speed of light is the same for A, B, and C. This problem was the starting point for the acute analysis that led Einstein to the discovery of relativity, as we shall see later.

Einstein, in addition to the "lightning" intuition of the equivalence between mass and energy ($E = mc^2$) - probably shared with Olinto de Pretto (see Chapter 4) - had another "lightning" intuition when he hypothesized that nothing can have a speed greater than c. But the question is: is speed c measured with respect to which observer?

Einstein, as we will see below (Chapter 16), deluded himself into thinking that he could construct the laws of physics regardless of the knowledge of an *absolute* natural reference system. However, as we will see, this illusion led to highly problematic consequences.

Other scholars instead referred to the simplest hypothesis, i.e., they hypothesized the existence of an absolute reference system, made up of a medium that they called "*æther*", and thus hypothesized that light moves at speed c with respect to "*æther*".

Following Einstein's discussion - and following some incorrectly interpreted experiments - the belief came to prevail that the æther does not exist. For the scheme described here, the æther exists, as we will explain below.

In any case, the success of the setup given by Maxwell's equations seemed perfect. In fact, it was not, as we will see below.

Chapter 6

Monads

"I advise those who wish to learn the art of scientific prediction not to spend their time upon abstract reasoning, but to decipher the secret language of Nature from the documents found in Nature: experimental facts."
[Max Born (1882-1970), *Experiment and Physical Theory*, 1943]

....

Everything (or almost all) of what has been said up to now is treated - largely and in greater detail - in school and university physics books. Instead, what follows here is not found in these books.

Let us remember, however, that physics is made up of "intuitions" which must then be confirmed by experience. If some "intuitions" lead to paradoxes, evidently the basic logical approach contains some logical "bugs" *ab origine*. If one replaces the original "intuitions" with other alternative "intuitions", one must never expect that the new "intuitions" can derive from the previous ones - perhaps thanks to a mathematical proof, etc. Rather, new "intuitions" must be temporarily accepted as working hypotheses, as new axioms, and as such require no formal demonstration. They just have to be accepted, and it is necessary to check whether they are suitable for explaining observations, especially those that - with the previous formulation - appeared paradoxical.

On the other hand, the interpretative scheme we propose in this journey of ours was never originally proposed as the "correct" one. Indeed, an insistent invitation has always been made to anyone who can propose other alternative schemes. The important thing is that any scheme must never lead to paradoxes and must be in accordance with all available observations.

Any discrepancies between the different proposals can, when possible, always be resolved only with reference to crucial experiments capable of resolving the controversies.

Therefore, the reader should prepare to also read about "intuitions" that necessarily clash with his most deeply rooted beliefs and intuitions.

Poynting's vector paradox

The Poynting vector highlights a non-negligible paradox, connected to the finite speed of the electromagnetic signal propagation. Let us try to briefly describe it.

Let's consider Fig. 6.1 and a thought experiment (Gedankenexperiment). Two circuits carrying an electric current jointly generate an electromagnetic "field" and deposit a specific density of electromagnetic energy at every point through "infinite" space.

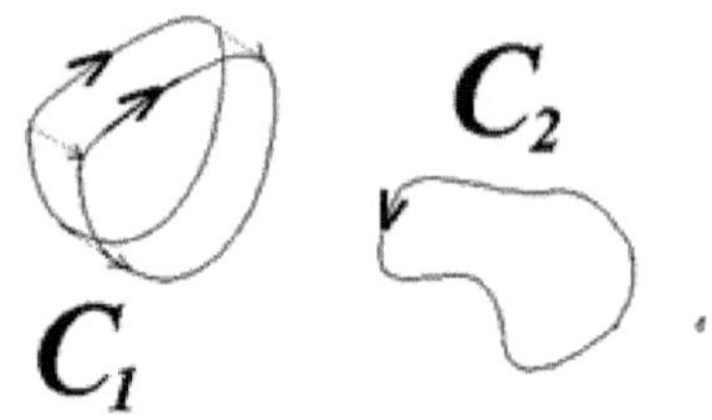

Fig. 6.1 – Two circuits C_1 and C_2, with an assigned although arbitrary shape, generate an electromagnetic field in space. In a first conceptual experiment, either one or the other of the two circuits, or even both, is moved and/or modified. The electromagnetic field in space changes consequently. In the second part of the conceptual experiment, the shape, orientation, current intensity, and position of the two circuits are restored, exactly as they were before the change. See text.

As a first step, let's move or modify one of the two circuits slightly. The "field" must change. Therefore, a radiation starts immediately, an electromagnetic wave - as described by the Poynting vector - which, in every direction, deposits or subtracts, according to need, an appropriate quantity of energy at every point to obtain the prescribed local density of electromagnetic energy. So far, there is nothing to complain about.

After some time - let's say after a time interval Δt - we restore the two circuits to the identical starting position, shape and orientation, with the same geometric configuration and the same current intensity, exactly as they were before the first part of the experiment. The new Poynting vector must retrace the entire space to restore the correct original distribution of energy density, point by point.

However, the first part of the thought experiment had involved a Poynting vector that, traveling at the speed of light c, had traveled a maximum distance $c\Delta t$. Therefore, an observer placed at a distance from the two circuits greater than $c\Delta t$ will not have noticed anything and will thus be completely unaware of what happened! This is a new fact - which had never been noticed in gravitational phenomena, conceived according to the hypothesis of action at a distance conceived by Newton, and according to the presumed "instantaneous" propagation of the gravitational signal.

That is, the system composed of the two circuits is associated with a spatial volume and a temporal interval (in the language of mathematics, it is said that a space-time "*domain*" is associated) such that nothing transpires outside the "*domain*" of what happened inside. For this reason, we call this space-temporal "*domain*" a "*monad*" (Fig. 6.2). Indeed, this conceptual experiment will be clarified better after introducing the concept of *SEA* (see Chapter 14).

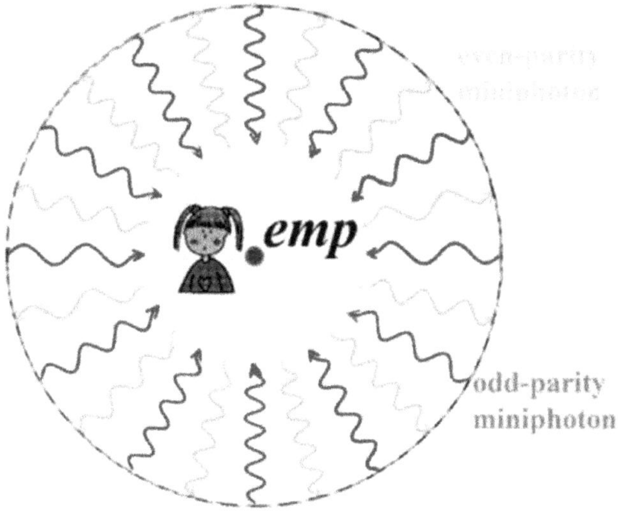

Fig. 6.2. Alice is the observer in the middle of the figure. Alice's "*luminous sphere*" denotes the spatial extension that an electromagnetic wave reaches during a time span Δt. This is a sphere of radius $R = c\Delta t$, where c is the speed of light. The figure shows the definition of "*monad*" in which Alice is immersed (see text). The Poynting vector that propagates in 3D space at the speed of light c is indicated by a blue (or light grey) wave moving outward (see Chapter 5). The returning Poynting vector is shown by a red (or darker grey) wave moving inward. The blue (or light grey) wave is composed of "miniphotons" of even parity, and the red (or darker grey) wave of "miniphotons" of odd parity, i.e., "antiminiphotons", i.e., with the arrow of time inverted. Every emp element always releases - as an endemic need - a sort of "*rubber band*" (or a "*double helix*") composed of photons of both parities. This figure will be discussed in better detail in Chapter 14 and Chapter 22. See text.

The term "*monad*" has been – and still is – well known in philosophy. It was used extensively, e.g., by Gottfried Wilhelm Leibniz, or Leibnitz (1646-1716), German mathematician, philosopher, etc., polymath. The concept of monad has, however, been used by many other philosophers since ancient times, even though always with partially different meanings from each other.

The Nature that we observe is always organized into entities structured as self-confined systems in space and time, which tend to limit communication with the outside. Man - like every observer - can be located either inside a monad or outside it.

For instance, the Solar System is a monad, and man has the possibility to observe the monad from inside. Thus, we can observe and define the laws that govern it. On the contrary, the "atom monad" can only be observed by man from outside, therefore, it is *impossible* for us to observe in detail what happens inside it. It may, however, happen that a monad expels some "*submonad*", i.e., a packet of emp – e.g., to avoid violating the "*principle of maximum density of emp*" (see Chapter 4). Thus, it happens that, with a process that we can call the "*slingshot effect*", the monad expels a "*sub-monad*" just like a bullet.

An observer who is outside the monad can thus know what is happening inside it only through the messages that the monad releases externally, or only through the "*sub-monads*" that are expelled like bullets due to the "slingshot effect".

"Slingshot effect"

Consider the monad atom. Man cannot see in detail what happens inside the monad, but he observes the "sub-monads", which he calls "*photons*", that are ejected from the monad.

On the basis of the photons observed, man can only try to assume what mechanism or process within the atom monad can give rise to these "photons". Furthermore, the empirical constraint is such that the mechanism can only be imagined by means of speculation, and therefore one can also imagine different equally effective models, without, however, any guarantee that one model is better than the others.

All of this is well known in atomic physics, which can be studied with the famous "*Schrödinger equation*" - by Erwin Schrödinger (1887-1961), Nobel Prize winner in 1933 for physics, Austrian physicist, naturalized Irishman.

The Schrödinger equation is certainly very effective and widely used - even if today its interpretation is still controversial and the subject of heated debates. We will come back to this topic. For now, it is sufficient to point out that Schrödinger's equation should be considered a "statistical" approach to the study of what happens inside the atom monad.

We only see the photons that are ejected by the slingshot effect, and we are only able to organize our observations "statistically". However, the "statistical" laws within every monad are very different from the "statistical" laws we apply in dealing the lottery, or playing cards, or for calculating the exit polls of a ballot, etc. Inside a monad nothing happens by chance, but only due to causes and processes which, due to the empirical constraint, we cannot fully know in detail. We will return to these concepts in detail later (see Chapter 10).

In general, it is reasonable to suppose that when, within any given monad, some process leads to physically incompatible conditions - e.g., when the "*principle of maximum emp density*" would be violated (see Chapter 4) – the monad expels the excess emp as a packet, itself being a "*sub-monad*".

Note that - as already noted previously (see Chapter 4) - the "*principle of maximum emp density*" prevents the formation of the mysterious and controversial "*black hole*". On the contrary, when due to gravitational implosion the concentration of emp violates this principle, some mechanism - possibly unknown to us - determines the violent expulsion of one or more sub-monads due to the slingshot effect.

The parity of the wave (d'Alembert) – Two specular spirals

We begin by considering the vibration of a violin string. The vibration of the string - caused by the friction of the bow that slides on the string - causes a transverse deformation of the string. The string is elastic. The deformation propagates from the point of contact of the bow along the entire length of the string and then is reflected at one end of the string and along the string, in the opposite direction. The pressure of the musician's finger changes the position - along the string - of the point where the reflection occurs, thus he changes the wavelength of the sound vibration. The vibration of the string hits the air and generates a sound wave that we perceive in the ear like a musical note.

The equation that governs the phenomenon is relatively simple and was solved for the first time by Jean-Baptiste Le Rond d'Alembert (1717-1783), French mathematician, physicist, astronomer, philosopher and encyclopedist, among the most important protagonists of the Enlightenment, co-director of the *Encyclopédie* from 1747 together with Denis Diderot (1713-1784) French philosopher, encyclopedist, writer and art critic, one of the most representative intellectuals of the 18th century.

The equation is called the d'Alembert equation, and formally has two simple solutions which were interpreted as elastic waves propagating along the violin string in opposite directions to each other.

Even in the calculation of an electromagnetic wave, the same d'Alembert equation must be solved mathematically, even if the meaning is very different from that due to the elasticity of a vibrating string. For this reason, someone hypothesized, by analogy, a sort of "elasticity" (?) of the æther, but this seemed like a very *ad hoc* hypothesis, however suggestive, and this led one astray. In any case, it was thought that the two solutions to the equation referred to electromagnetic waves propagating in the opposite direction. Conversely, the meaning is much subtler and more important than a simplistic analogy to an elastic mechanical medium.

Let's reconsider Fig. 5.5, which is enlarged as in Fig. 6.3. The case of propagation - as in Fig. 5.5 - is said to represent "*polarized*" light, where the "polarization plane" is defined with the plane of $\boldsymbol{B}$. There are materials that are transparent only for light polarized in a certain plane, and they are opaque if the plane of polarization is rotated 90°.

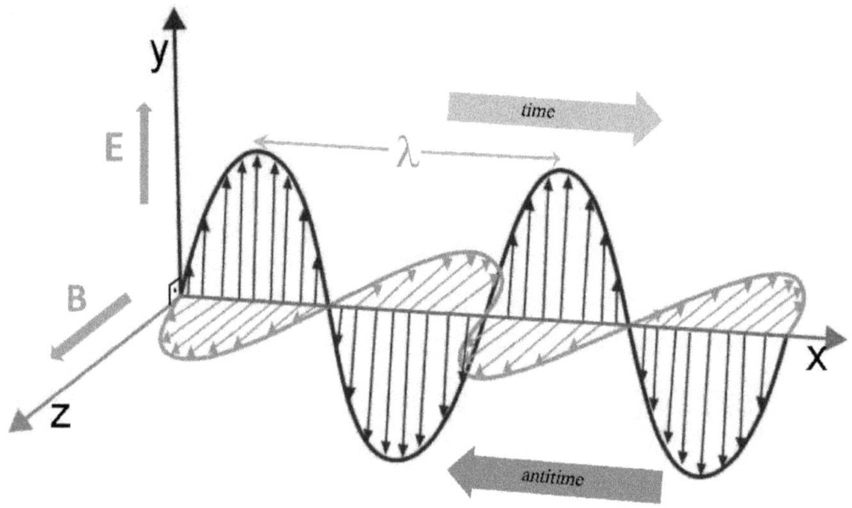

Fig. 6.3. The two solutions of the d'Alembert equation correspond to the propagation of an electromagnetic wave with mirror parity relative to each other. A wave propagates in the world α with time. The mirror wave propagates in the γ world with reference to antitime. See text.

For instance, for many years, there have been polarizing sunglasses on the market that reduce the intensity of light, excluding the components that do not have the correct polarization allowed by the lens. In fact, a generic light ray contains both radiations of different wavelengths and of different polarization. An appropriate medium is transparent to only some radiations.

Consider Fig. 6.3. It is the same as Fig. 5.5 which can be seen as the overlap of two signals. One signal propagates in time. The other signal propagates in antitime, where everything remains identical, except for the mutual orientation between $\boldsymbol{E}$, $\boldsymbol{B}$, and the direction of propagation which must be reversed. The propagation of the two signals in time and antitime therefore represents two specular phenomena, just as the right hand is mirrored to the left, and the glove for the right hand is mirrored - but not identical - to that for the left hand.

We now come back to the solution of d'Alembert's equation. With a little amount of simple algebra - which it is not necessary to remember here, despite being almost elementary, but we do not want to burden our journey with unnecessary formalisms - it is possible to recognize two solutions characterized by circular polarization, rather than planar polarization.

That is, the vectors $\boldsymbol{E}$ and $\boldsymbol{B}$, although remaining mutually orthogonal, rotate like a spiral staircase. In the case of one solution, the rotation is in the direction of the clock hands (i.e., clockwise), and in the case of the other solution, in the opposite direction, i.e., counterclockwise. Remember, e.g., the case of the access and exit stairs from the *Vatican Museums* in Rome, which are two spirals linked to each other, but distinct, and traveled by visitors entering and exiting in opposite directions. The St. Patrick's Well in Orvieto also has two linked spirals, which were used for those who went down to get water with empty containers and those who came up to take them up.

It follows that a ray of light - of a single wavelength, i.e., of a single color - propagates like two superimposed spirals as in Fig. 6.4.

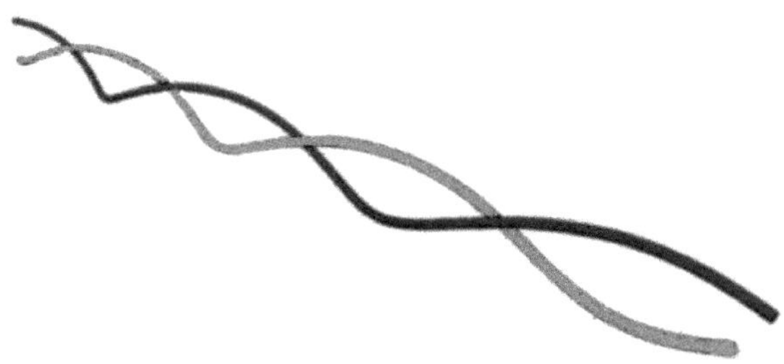

Fig. 6.4. If the two specular electromagnetic waves, which correspond to the two solutions of the d'Alembert equation - transformed by the polarization according to orthogonal planes into circular polarization (see text) - are superimposed, we have a structure with two linked spirals. One spiral travels in one direction (that of "time", here in blue or light grey), the other spiral in the opposite direction (that of antitime, here in red or dark grey). This structure is typical of a photon which - having a component that flows with time, and a component that flows with antitime - is outside of time and antitime, i.e., it is in the β world. This double spiral structure recalls the famous "St. Patrick's Well" in Orvieto, or the visitor access/exit staircase at the *Vatican Museums* in Rome. See text.

Note that Fig. 6.4 represents two continuous spirals. In fact, we will see later that these spirals should be considered as made up of small discrete elements, a bit like the staircase of the *Vatican Museums* or that of St. Patrick's Well, which are made up of discrete steps.

α, β, γ worlds

Thus, we will claim that the photon is composed of two spirals linked together and superimposed. The photon is then outside of time, as it has a component (spiral) which flows in time, and a component (spiral) which flows in antitime. A photon is then in the β-world of Fig. 4.1, here reproduced as Fig. 6.5. Note that, as already stressed, in the β world - since either "time" or "antitime" does not exist - it is not possible to define a "velocity" or an "antivelocity".

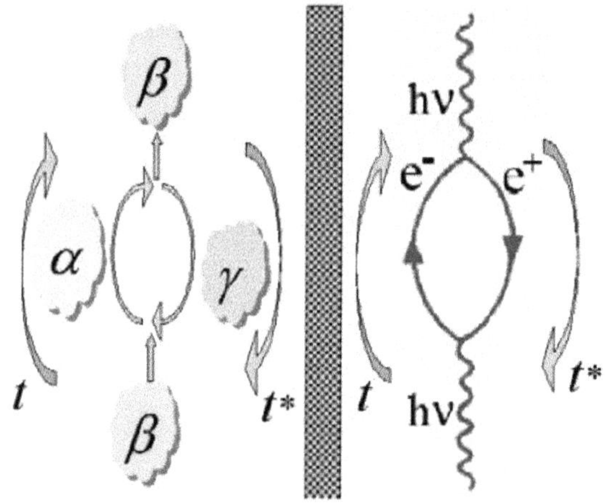

Fig. 6.5. In the right panel, a photon with energy $h\nu$ is indicated, which materializes in a pair of an electron (e^-) and a positron e^+ (more properly called antielectron). The phenomenon is paradoxical and unexplained, and is observed regularly, but the phenomenon also concerns other particles, other than electrons and antielectrons. This is a mystery that we must explain. All of this can be interpreted (left panel) with the existence of three worlds. The "absolute" world is the β world that is timeless. Occasionally, from the world β, objects are generated in two symmetrical and mirrored worlds α and γ. In world α time flows as we know it. In the γ world, time flows in the opposite direction, from the future to the past, i.e., we have an antitime. This logical scheme seems to be the only possible one to solve the mystery of antimatter. The concept of a timeless "absolute" world is not new, but it is found in *Genesis* and in sacred Hindu texts (see Chapter 9).

When the two spiral components decouple - and each spiral flows in space independent of the other - there is the phenomenon of the "materialization" of the photon in a pair of particle and antiparticle, as represented schematically in Fig. 4.1 or 6.5. The particle that flows in "time" is in the α world, the world of matter. The antiparticle that flows in antitime is found in the γ world of antimatter.

In the world α the photon is always observed moving at the speed c. In the γ world, the photon is observed moving at speed c but referred to the *antitime*.

Particles and antiparticles, respectively in the α and γ worlds, are emp that always move at a speed lower than that of light. It is as if the emp contained in the photon, by "materializing", became emp in some way "*degraded*", given that it moves at a lower speed, almost as if it were slowed down by its intrinsic nature.

This is a very surprising and indeed bizarre first result. Matter and antimatter, time and antitime, and worlds α, β, γ are connected to the solution of d'Alembert's equation! Is this a coincidence? A coincidence would truly be a rarity, since nothing in Nature happens by chance. The paradox of matter and antimatter now seems much less mysterious. It is written in d'Alembert's equation!

In any case, those who proposed this interpretation invited others to propose alternative interpretations – and every new proposal will always be considered with great attention as it should be. However, as long as there is no other explanation of the "mystery" of antimatter, the formulation described here is the only interpretation that is available today.

The path of science occurs through ideas proposed and discussed, even controversial ideas, although always on the basis of objective concepts, certainly not on arguments like "*everyone says so...*" which simply cannot be part of science!

Conclusion

We have thus reached an important stage in our journey. But we are still burdened by the concept of *continuity*, which is intrinsic in the formulation of Newtonian dynamics, altogether with the concepts of "force", mass, energy, and Maxwell's concept of "field". That is, we are treating phenomena as when phenomena are smoothed and interpolated by continuous functions, i.e., as when we have replaced the discrete data of Fig. 2.1 or 2.2 with the continuous data of Fig. 2.4. We made reference to averaged, smoothed concepts, as is done when studying "statistical" trends. Now there remains the next step, aimed at resolving the famous and celebrated dichotomy of light as a wave or as a quantum (photon).

Before tackling this difficult - well-known and classic - dilemma, it is necessary to clarify the concept of time and its mysterious flow, only in one direction - associated with the no less mysterious aging of everything that exists in Nature without exception whatsoever.

However, for this purpose, it is essential to first consider the golden ratio.

Chapter 7

The Golden Ratio – Fractals

"Provare a liberare un popolo che vuole rimanere servile è difficile e pericoloso quanto provare a schiavizzare un popolo che vuole rimanere libero." [*"Trying to free people who want to remain servile is difficult and dangerous as trying to enslave people who want to remain free."*]

Niccolò Machiavelli (1469-1527)

....

The golden ratio

Let us now venture into a much debated - and fascinating – topic, which until very recently seemed mysterious - and today in general it is still considered a "mystery", which has certainly been observed in Nature in a myriad of forms, but always inexplicable.

The ideal proportions of the human body show the "*golden ratio*" – sometimes denoted by other equivalent names, such as "*golden section*" or "*golden number*" or "*Phidias' constant*" or "*divine proportion*" or "*Golden Rule*" - and it also appears in several parameters of the Solar System, e.g., in sunspots, or in many parameters of the orbits of the planets.

The golden ratio also appears in electronic circuits, in the shape of leaves, in general, in the biosphere, in architecture, in paintings (particularly in Renaissance paintings), in musical harmony, in stock market transactions, etc. The golden ratio seems to reflect a mysterious search for regularity throughout the natural world, as we will highlight below. The most common case known to all is the shape of a credit card whose dimensions (ratio between the major side and the minor side of the card itself) is precisely the golden ratio.

Mathematics, geometry, all natural sciences, including psychological and artistic phenomena, music, etc., have been the subject of research explicitly connected to the "mysterious" golden ratio. Only very few elements can be briefly outlined here in an essential and easily understandable way, useful for our logical journey, but the literature on this topic seems truly endless.

We begin with a historical note. Leonardo Bonacci (c.1170 - c.1250), better known as Fibonacci - and also as Leonardo of Pisa, Leonardo Pisano Bigollo, Leonardo Fibonacci - was an Italian mathematician. He is considered one of the most talented western mathematicians of the Middle Ages. His main concern was focused on economic applications, although his book "*Liber abaci*" (Book of the Abacus) denotes acute mathematical thought. He must be considered in the context of the controversy between the "*abacists*" and the "*algorithmists*". The "*abacists*" strongly supported the use of the abacus and - in a way consistent with the Roman monetary system - preferred to use fractions expressed in $1/12$ of a unit (even today, dozens are often used, or the subdivision of the hours of the day, or an angle in degrees, etc.). In contrast, the "*algorithmists*" referred to the use of algorithms that used multiples or submultiples of ten, such as the decimal system used today in science, after the enlightened Napoleonic reforms. Fibonacci was very important as he popularized the Indo-Arabic numeral system in the Western world.

Let us now consider the so-called "Fibonacci rabbit problem", which he described in his "*Liber abaci*" and which will lead us to the famous sequence of "*Fibonacci numbers*". The name "*Fibonacci numbers*" seems to have been first used by the French mathematician François Édouard Anatole Lucas (1842-1891), but the "*golden ratio*" was certainly known since remote antiquity, e.g., by the builders of the great Egyptian pyramids, or by the great Phidias (c. 490 BC – c. 430 BC, Athenian sculptor and architect, active from approximately 470 BC) when he designed and built the Parthenon. Today, the golden ratio is denoted by the letter φ (phi) in honor of Phidias.

Fibonacci was concerned with economic aspects. His argument was as follows. Rabbits should be considered as male and female couples, and every couple generates a new couple every month. Once a new couple is two months old, the new couple generates, in turn, each month a new couple. Indeed, Fibonacci made an arbitrary exemplification - although useful for his mathematical argument. In fact, a couple of rabbits procreate a large number of rabbits at a time, and sexual maturity requires at least $6-8\ months$.

In Fibonacci's calculation, starting with a newly born couple at the beginning of a year, how many couples of rabbits will there be at the end of the year? If we make a table, we formally obtain the sequence of Fibonacci numbers, i.e., we find that every number (positive natural integer) of the sequence is defined as the sum of the last two previous numbers of the sequence. The mathematical sequence never ends

$$0, 1, 1, 2, 3, 5, 8, 13, 21, 34, 55, 89, 144, 233, 377, 610, 987, 1597, \ldots$$

We call F_j the Fibonacci number, which, in the sequence, is found in the progressive position j. Let's consider the ratio between every number in the sequence and the number that precedes it. That is, in symbols, we consider the ratio F_{j+1}/F_j. This ratio - as the order of the number j in the sequence increases - gets increasingly closer to the golden ratio φ, however, without ever reaching it. In mathematical language, it is said that the limit of the relationship F_{n+1}/F_n for n tending to infinity (∞) is precisely φ, i.e., in symbols

$$\lim_{n\to\infty} \frac{F_{n+1}}{F_n} = \varphi \tag{7.1}$$

The bibliography on the golden ratio - with multiple mathematical, geometric, and other facets - seems endless. Nor can we try to summarize what is impossible to summarize. Let's make just a few comments useful for our purpose.

We call "*natural system*" every composed system, e.g., several couples of rabbits. Every system can also be split into two smaller systems. For instance, a system of *10* couples of rabbits can be divided into two systems, one with *7* couples of rabbits and the other with *3* couples of rabbits, etc. Let's assume that every system, more or less large, has similar behavior to all other systems. We call this property "*additivity*" – i.e., every natural system satisfies the property of additivity, as the sum of two systems always behaves like every individual system.

In general, a system can be decomposed into one or more smaller systems. If this is possible, the system will be said to be "*composite*". If, however, a system cannot be broken down into smaller systems - e.g., a couple of rabbits cannot be broken down since it is essential to have a male and a female - then the system is called "*minimal*".

Analyzing "all" the known cases in which the golden ratio is observed, we see that the golden ratio is always found for every system of any nature that satisfies *additivity* and which, furthermore, satisfies some *variational principle*, i.e., the natural system always seeks, in any case, either the maximum or minimum of a suitable quantity. For instance, one or more couples of rabbits always seek the maximum possible reproduction, or even one or more stock traders behave in a similar way (additivity), and always seek the maximum income, etc.

Let us consider two natural *systems,* which for brevity we will call A and B. Let us assume, as a working hypothesis, that each of them can be associated with a number, which we call a and b, respectively, and we also assume that the *composite system* A+B is associated with the number $a + b$. Let's say that these assumptions are part of the *additivity* hypothesis of systems.

We assume that every system evolves in *search of an equilibrium,* and - in the case $a > b$ - equilibrium will occur only when

$$\frac{a+b}{a} = \frac{a}{b} \tag{7.2}$$

and also

$$\frac{a}{a+b}=\frac{b}{a} \tag{7.3}$$

Please note that this is an assumption - a working hypothesis - that will help us resolve the "mystery" of the meaning of the golden ratio. According to mathematics of lower secondary school, we see that dividing by b both the numerator and the denominator of the first member of (7.2) one gets

$$\frac{(a/b)+1}{(a/b)}=\left(\frac{a}{b}\right) \tag{7.2'}$$

and similarly, dividing by a the numerator and denominator of the first term of (7.3) we have

$$\frac{1}{1+(b/a)}=\left(\frac{b}{a}\right) \tag{7.3'}$$

Equations (7.2') and (7.3') each have only one unknown, that is, (a/b) and (b/a), respectively. They are second degree equations, i.e., elementary equations, and the solution can thus be calculated

$$\frac{a+b}{a}=\frac{a}{b}=\varphi=\frac{1+\sqrt{5}}{2}=1.61803398874989484820\ldots \tag{7.4}$$

$$\frac{a}{b}=-\Phi=\frac{1-\sqrt{5}}{2}=-0.61803398874989484820\ldots \tag{7.5}$$

where φ is called "*golden ratio*" and Φ "*conjugated golden ratio*".

In fact, notice that, if a and b are two consecutive Fibonacci numbers, with $b<a$, the next number is $a+b$. So (7.2) requires that the relative ratios must be identical. Note that both φ and Φ are irrational numbers, i.e., they have an infinite number of decimals (see Chapter 2). This property, as will be seen, is of *fundamental importance*. Finally, it is shown that these numbers φ and Φ have the following properties, which seem almost "magical" in their simplicity

$$\Phi=\frac{1}{\varphi}=\varphi-1 \qquad \frac{1}{\Phi}=\Phi+1 \tag{7.6}$$

$$\varphi^2=\varphi+1 \qquad -\Phi^2=\Phi-1 \tag{7.7}$$

The assumption (7.2) or (7.3) can be considered the expression of the search for equilibrium by a natural system, equivalent to the property that, in general in physics - according to the principles of Newton's and Maxwell's dynamics - manifests itself as the search for a minimum amount of energy. That is, the search for a minimum energy is what is most frequently called the "*variational principle*".

In other words, Nature seeks every variation that aims to smooth out every extreme, Nature seeks to smooth out differences, to reduce cases of extreme violence. Nature always seeks a balance that avoids catastrophic phenomena.

In fact, as a matter of fact and as already mentioned, all known cases of phenomena - in which the existence of the golden ratio has been found - always satisfy two fundamental requirements: *additivity* and a *variational principle* of some kind. Furthermore, in any case, the criterion is always to seek the maximum profit - and in fact, stock market transactions curiously and mysteriously satisfy the golden ratio. Surprisingly, by connecting several identical electrical circuits, a new composite circuit is found (*additivity*), where the current circulates similarly to every simple circuit, and the composite circuit always tries to minimize the energy of the system (*variational principle*), and we observe the golden ratio, etc.

The golden ratio is found in the "canons" of Renaissance painting - as if it were a "word-of-mouth" between the "workshops" of the various painters. This is well known, and a very extensive literature exists. These great painters and sculptors certainly had no mathematical motivations for doing this. They only responded to an "instinct" of beauty, as we will clarify in Chapter 8.

Likewise, in terms of an analogy, for centuries the science of "perspective" was the - often unattainable - dream of every painter, even very famous ones. Some painters achieved absolutely impressive results – e.g., Piero della Francesca (1412/1416 – 1492), or Piero di Benedetto de' Franceschi, painter and mathematician among the most emblematic personalities of the Italian Renaissance, "*el monarcha de la pittura*", exponent of the second generation of painter-humanists.

These excellent artists applied a logic for "perspective" but probably could not explain it. No one knew how to define it as a discipline of geometry. The lives of many excellent artists show that - thanks to their great personal skill and thumb rules - they achieved unsurpassed results. They considered it all as professional know-how, to be kept secret within their "workshop". This attitude persisted until the full formal and logical definition of "*perspective*" obtained - much later - by mathematicians by means of "*projective geometry*". Analogously, these humanist painters conceived the golden ratio as an essential ingredient for the canons of beauty in a painting, a "secret" of the "workshop".

In summary, to explain the "mystery" of the golden ratio we can synthesize the set of assumptions described so far with the following "*Golden additivity theorem*" which can be stated as follows.

If a natural system satisfies additivity, and - moreover - if any two systems A and B can be associated with two scalars (i.e., simple numbers) a and b, which are respectively, successive integer numbers in the Fibonacci sequence, such that the system satisfies a variational principle expressed as

a/b, this condition is sufficient for the system to comply with the "Golden Rule". Equilibrium is reached every time (for $b < a$) it is $a/b = \varphi = 1/\Phi$. However, the natural system will never reach the true final equilibrium, because φ is irrational.

The last part of this statement is linked to the fact that the numbers φ and Φ are irrational, as better clarified below.

The search for equilibrium

Let's make a hypothesis - very important for our journey - which appears to be in agreement with observations. Every system would seem to seek equilibrium through a succession of states of the system, which could be associated with an appropriate series of numbers that are extracted from the sequence of Fibonacci numbers. In fact, since the Fibonacci sequence is infinite, the natural system never reaches a final equilibrium.

The natural system is therefore increasingly approaching ideal equilibrium, without, however, ever being able to reach it. This assumption, as we will see in Chapter 9, has consequences of enormous importance for getting to the bottom of any paradox.

During the evolution of the natural system, some environmental conditions change where the system is located, and the final objective of the ideal balance to be achieved consequently changes. For example, in the case of Fibonacci's rabbits, it is necessary to consider the availability of sufficient food, as well as of living space, the absence of epidemic phenomena, etc. As environmental conditions change locally, the system changes its course, evolving towards the new objective of the final equilibrium that it must try to achieve.

In this mechanism, a fundamental role is played by the fact that φ is an irrational number. Indeed, the inability of the natural system to achieve equilibrium is fundamental. In fact, suppose - absurdly - that the natural system actually manages to reach the final ideal equilibrium. In this case, the Universe would be in total equilibrium, it would be an enormous immobile swamp, where nothing changes - and where it would be impossible to even perceive the flow of "time" (as we will see more fully in Chapter 9). There would be no life whatsoever. In other words, everything that we perceive in the world where we live - as we know it in a better or worse perspective - depends *critically* on the *irrationality* of φ.

It is, therefore, natural to ask ourselves the question of whether *all* the numbers in the Fibonacci sequence can be associated with the evolution of some given natural system. The alternative is that, on the contrary, there are different subsets of the Fibonacci numbers that are eventually used by different similar natural systems. For instance, we know that there are several types of pears. They can all be obtained from the same tree, the pear tree, as long as they are properly grafted. Every different grafting makes different branches of the same tree to provide different types of pears. That is, the biological organism (the pear tree) is the same, but every branch follows different rules, which is why it produces different types of fruit. The question is if different branches refer to different subsets of the Fibonacci numbers.

Another example is shown from Fig. 7.1. Walnuts, grown on the same tree, more commonly show a shell divided into two mirror-image halves, with symmetrical and mirror-image kernels, divided by an internal kernel. In rare cases, however, a walnut shows the shell and internal kernels divided into three rather than two sectors (even if sometimes only the external shell is divided into three equal parts). We wonder if the phenomenon can be associated with different "families" or subsets of Fibonacci numbers.

Fig. 7.1. Two walnuts, growing on the same tree, may have the common two-part structure, but occasionally they may have a three-part structure. See text.

Are there subsets of Fibonacci numbers recognizable in mathematics?

A first selected class of Fibonacci numbers is made up of numbers that are also prime numbers. Some "minimal" natural systems could be related to prime numbers. In fact, some primes are also Fibonacci numbers. Many Fibonacci prime numbers are known, but it is not known whether there are infinite numbers of them. As a further curiosity, it is hypothesized that if a Fibonacci number F_n is prime, then n is prime. But the opposite is not necessarily true, given that, e.g., the 19th Fibonacci number is $4181 = 37 \times 113$.

Another possible selected class of Fibonacci numbers consists of numbers that satisfy the "*Goldbach conjecture*".

The Prussian mathematician Christian Goldbach (1690-1764) wrote a letter in 1742 to the famous Swiss mathematician, physicist, and astronomer Leonhard Euler (1707-1783) and proposed the following conjecture: "*every integer greater than* 5 *can be written as the sum of three prime numbers.*"

Euler responded by reformulating the problem in the following equivalent version: "*every even number greater than* 2 *can be written as a sum of two prime numbers.*" The form in which the conjecture is currently formulated is Euler's and is sometimes also called Goldbach's *strong conjecture*. By contrast, Goldbach's *weak conjecture,* which is implicit in the strong conjecture, states that all odd numbers greater than *5* can be written as the sum of three primes.

In fact, a property that has not yet been fully proven is called a "*conjecture*". That is, it has been verified that a certain property holds up to very large numbers, but a proof that holds for all numbers, up to infinity, is missing. It has also been shown (by Lord Christopher W. Monckton, Viscount Brenchley, Peer of Scotland) that a particular property of the numbers in the sequence that satisfy Goldbach's conjecture tends to the limit of the golden ratio.

That is, a possible hypothesis (*Monckton's conjecture*) is that a given sequence of Fibonacci numbers - which satisfy the Goldbach conjecture - could be associated with particular natural systems.

In other words, nothing is certain. The meaning of the golden ratio - as widely found and widespread in Nature as it is "mysterious" - *could have very important consequences for the explanation of many natural phenomena.*

Keep in mind that science is *always* made up of hypotheses, which must then be verified. Science must deal with the frustrating limitations of the empirical constraint, and we must "get our hands dirty" with the reality of what is observable - and not more than that. Natural science is not a logical embroidery like mathematics - which is a real challenge of the human mind, a marvelous monument built with logic, for which, given some premises, one wonders what all the conclusions that follow are. In natural science, discovering that a working hypothesis is incorrect always helps to find another hypothesis that is more in agreement with observations. That is natural science.

We will see in Chapter 9 how these comments - apparently bizarre - play an important role in clarifying the very meaning of "time" which has been the subject of heated discussions since prehistoric times - and is in fact strictly essential for our logical journey aimed at coming to terms with paradoxes that have now plagued physics for over a century.

Fractality

But we are not finished with the "surprises" yet. Consider Fig. 7.2. It is a Roman broccoli, now accepted in international literature as a classic example of a structure that repeats itself from the smallest to the largest scale. In fact, if you take a photograph of a detail and hide its outline, you don't understand whether you are referring to a larger or smaller part of the broccoli. The system is said to enjoy the property called *self-similarity*. That is, the same shape is repeated in shapes that are geometrically similar but of different dimension. The Universe is full of phenomena that display this beautiful property.

Fig. 7.2. Roman broccoli (*Brassica oleracea*) in the garden, a single plant, and a detail of the plant. The plant repeats the same structure on different scales through self-similarity. See text.

This concept led to the definition of what is today called "*fractal*". The discovery is comparatively very recent and dates back to 1975 when Benoît Mandelbrot (1924-2010), a great Polish mathematician naturalized French, published the volume "*Les objets fractals: forme, hasard et dimension*". Some properties of fractals had previously been studied by other scholars, but considered as isolated curiosities, without using the name "fractals". Today the literature on fractals is enormous. Studying their mathematical properties generates images of rare æsthetic beauty, often seen every day in advertisements etc.

Another classic and very simple example is a detailed map of a rugged coast. If you don't indicate a spatial scale on the map, we don't know if it is, e.g., the map of a large gulf or the map of a tiny cove. But there are countless application cases, from microphysics to the study of galaxies.

For instance, with fractal analysis, we can recognize the type of "whine" - measured with special acoustic sensors - released by a solid object (metallic, rock, or other) when subjected to stress. The effort causes an "aging" of the solid structure, and the type of "lament" changes with ageing, almost as if it were a different musical instrument. With this, it is possible to know whether the object (an artefact or a structure of the Earth's crust, or a volcanic building) is more or less close to the fracture, causing a possible catastrophe etc.

Obviously, in Nature, the fractal property is valid within certain limits. In fact, for a broccoli, it is not possible to go beyond the maximum size of the vegetable, or below the size of a molecule, etc. In mathematics, however, fractals are ideal objects, with an unlimited scale from zero to infinity, which allow the construction of incredibly beautiful images... We will not spend time explaining the mathematics of quantitative methods for these applications (both graphic and æsthetic, and for practical purposes). Rather, we show how this property - which is very widespread in nature and which can be called "*fractality*" - is connected with the golden ratio.

That is, since the observation of the golden ratio property is very widespread in nature - as is the observation of the fractal property - the two properties are in fact two facets of a common property that is unequivocally and often commonly observed in Nature. It can be argued that it is reasonable to assume that, when one of the two properties is observed, the other is also still hidden, but one day it will be discovered by some scholar.

This is a very surprising and important result. Let's see the detail - and for this purpose, some very little knowledge of elementary Euclidean geometry is sufficient. Consider Fig. 7.3, (proposed in 2021 by Mercedes Orus-Lacort of *College Mathematics UOC–UNED*, Spain, and Christophe Jouis of *Université de la Sorbonne Nouvelle*, Paris, then reinterpreted in 2022 by G. P. Gregori and co-authors). The right triangles ABC and ADF are similar. Therefore, for this particular right triangle, we have that $(a + b)/a = a/b$ which in fact coincides with (7.2). Therefore, the same Fig. 7.2 says that a suitable right triangle is reproduced in Nature by similar triangles, as required by the principle of fractality. That is, the principle of fractality, with the appropriate relationships expressed in (7.2), necessarily involves the golden ratio.

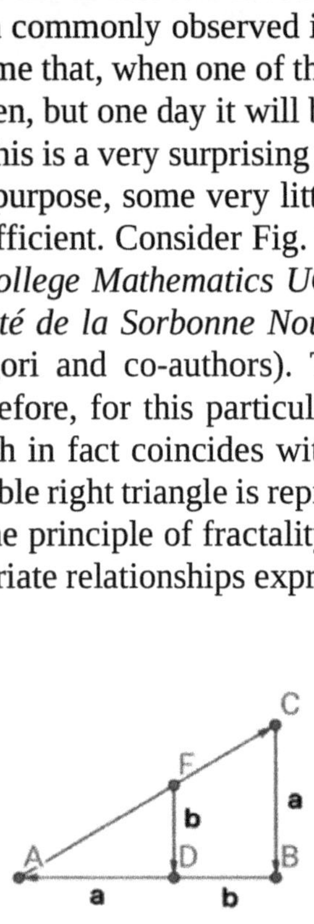

Fig. 7.3. A right-angled triangle to generate the golden ratio shows that the hypothesis of the golden ratio implies the hypothesis of self-similarity, therefore of fractality. See text.

In conclusion, fractals and the golden ratio are two facets of an intrinsic and widespread regularity in natural phenomena. If a system respects the Golden Rule, then it also respects fractality. If a system respects fractality, it would be necessary to consider in detail how the system also respects the Golden Rule. But these are minor "technical" aspects. All this certainly strengthens the credibility of the interpretation previously proposed for the "mysterious" golden ratio.

In nature, nothing happens by chance. Furthermore, all this - together with the fact that the golden ratio is an irrational number - is linked to the perpetual search for equilibrium of all natural systems. Equilibrium is never reached, however, and this - as we will see - is essential to prevent the world from being a swamp of death and desolation, and is also essential for our perception of time.

But let's proceed in order...

Chapter 8

Definition of Beauty

"Nature is only to be commanded by obeying her"
Francis Bacon, 1st Viscount St Alban, (1561-1626)
Novum Organum Scientiarum (1620)

....

There has long been discussion about the definition of beauty - e.g., comparing images that have intrinsic regularities related to the golden ratio, with other images without this property. Then, "democratically" a sample of observers was asked to state which images appear most beautiful to them. The experiment worked. But the mechanism that determined the sample observers' decision remains unknown. Indeed, the definition of beauty has a connection with the golden ratio in both visual and auditory arts.

A plausible explanation is that living beings want security – just as a survival instinct - so much so that human psychology seems to unconsciously perceive security as a feeling called "beauty".

Conversely, living creatures with no psychology (?) simply respond to the instinct of survival, and obviously no one knows what role some more or less sketchy form has, that we can assimilate to our "feeling" of "beauty".

The auditory arts, i.e., music, allow us to quantitatively analyze and support this notion. For example, qualitatively, a harmonious sound is pleasing to every ear, even those not trained in music, and furthermore it does not even disturb animals. Instead, the screech of a door that creaks or slams violently, or the sound of strong thunder, or that of an explosion, or a firework, can strike fear into every living being.

For readers who are unfamiliar with the nature of sound, we briefly introduce some simple, elementary notions.

Consider a violin string. As the string vibrates, it "whips" the air. Thus, compressed layers of air are generated, which are then decompressed immediately. The phenomenon propagates in the air a bit like the rings in a pond where a pebble falls. The compression/decompression reaches our ear, causes the eardrum to vibrate, and we thus perceive the sound.

As the length of the violin string varies - determined by the pressure of the musician's finger - the musical note changes, and every note is rigorously associated with a certain frequency, or wavelength, of the vibration. Therefore, every note is unequivocally associated with a number, usually expressed with the frequency ν (Greek letter "nu") which is equal to $1/T$ where T is the "period" of the sound.

A study by Richard Merrick - "*Interference. A grand scientific musical theory*" (various editions between 2009 and 2011, followed by further developments published in 2019) - examines the theory of so-called "*harmony*" since prehistoric times. The author's analysis - as he clearly states - does not concern physics or mathematics, but rather music.

Conservatories of music (i.e., music schools) teach which chords and sounds are allowed and which are prohibited, in order to obtain "*harmony*". These rules are the result of long-established conventions, based solely on æsthetic sensation. Merrick performed a quantitative analysis of the frequencies of permitted sounds. He thus discovered that harmony is achieved when the frequencies meet the golden ratio. The result is undoubtedly surprising and impressive.

A related 2021 study, by J. H. E. Cartwright, D. L. González, and O. Piro, re-establishes the connection with the *musica universalis* of the cosmos, of ancient Pythagorean memory.

That is, a sound - that respects the golden ratio in pairing notes etc.- is judged pleasant and reassuring, artistic by man, and is judged not to be worrying by the living world. On the contrary, any sound that does not respect the golden ratio is "unpredictable" and could be a harbinger of life-threatening hazard. It therefore seems scary or even terrifying.

This shows that the human brain (and also the more or less developed brain of every living being) is, in fact, a powerful analog computer that, in real time, almost immediately perceives whether a sound is harmonious (i.e., satisfies the golden ratio) or discordant.

Let's remember the difference between a "*digital*" and an "*analog*" electronic computer. A *digital* computer is based on objects expressed with numbers ("*digits*" in English), and solves equations thanks to appropriate programming, and finally provides a result. In contrast, an *analog* computer relies on electrical circuits that are governed by equations similar to those in the problem under consideration. By making appropriate measurements on correctly chosen knots of the circuit, one obtains - by analogy - a quicker, almost immediate, response to the solution sought.

Furthermore, the accuracy of the response of an analog computer - built by man - is generally less precise than that of a digital computer. However, this limitation is only indicative. In fact, every practical case requires a long and accurate specialist study.

The human brain (and also that of every living being) therefore functions as a highly reliable analog computer which provides - in real time, or practically instantaneously - the information that is essential for survival, recognizing what is regulated by the golden ratio, and what could instead be a harbinger of an unpredictable and potentially hazard evolution, or a possible serious threat to life.

Concerning the visual arts, note that any image that displays fractality – e.g., a Roman broccoli or a cauliflower (see Fig. 7.2) or some fanciful figures – is perceived as “beautiful”. That’s the *definition of beauty*.

Or in stock market transactions, every operation that respects the golden ratio is considered promising, otherwise, it is considered risky and to be avoided, etc.

Therefore, if we could translate some difficult problem into sounds, we should - in principle - be able to solve extremely complex calculations by exploiting the natural analog computer represented by the human ear and brain. Likewise, when looking at a painting, the human brain seems to perceive calm and security when the proportions meet the golden ratio. Thus, the observer judges the painting to be beautiful.

Various authors, e.g., have hypothesized a possible connection of the golden ratio with chaos theory. For instance, when discussing the discovery that the golden ratio φ is found in the time series of sunspots, it has been suggested that this fact can be interpreted as “a sign of chaos”.

In this same sense, consider a natural system that is quasi periodic. To be concrete, let’s consider two identical eggs, one boiled, the other not. The two eggs are placed horizontally on a plain table. Let’s push each of the two eggs. The hard-boiled egg will spin several times. This is an example of an almost periodic natural system. In fact, the motion is not periodic in the strict sense, since the friction with the table surface (and with the air) progressively dampens the rotation of the egg. It is thus an almost periodic phenomenon, and the phenomenon fades away with time.

In contrast, the other egg, which has not undergone any cooking, contains within it a viscous fluid composed of egg white and yolk. These fluid components tend to oppose and dampen the egg’s motion. The cause lies in the friction dissipated within the internal fluids that try to oppose the motion – and this happens due to a simple reason of inertia. The end result is that the egg that is not hard-boiled quickly becomes stuck.

More generally, we can say that every quasi-periodic system of dissipative systems (i.e., with internal frictions - and in fact every natural system always has more or less relevant internal frictions) evolves towards chaos, since the heat dissipated by friction is a clear step forward towards chaos.

This rule has value for the whole of physics – and represents a truly great and famous paradox, well known and widely debated. In fact, Nature on the one hand, tries to evolve towards states of balance and "quietness" that avoid "catastrophic" situations of any sort. On the other hand, every form of friction tends to smooth out and destroy every regular structure, pushing everything towards fragmentation, towards a sort of primordial "mixture", where all order disappears.

A mathematical form was given to this concept by Ludwig Eduard Boltzmann (1844-1906), an Austrian physicist, mathematician, and philosopher, who invented a function called "entropy". There is no need to explain this detail of mathematical language in this context. It is now a classic and well-established concept reported in every physics treatise. It is here sufficient to stress only the true and distressing paradox arising for physics. That is, if the entire Universe evolves by increasing entropy, what will happen when all ordered forms have been transformed into a shapeless ocean of chaos? Will the world end?

This paradox will also find an explanation and a logical solution during our journey, because entropy grows in world α while in world γ antientropy (i.e., the entropy of antimatter) grows. When world α and world γ are annihilated, entropy and antientropy compensate each other, and the global system restores the initial value of null entropy and antientropy. All of this will be discussed in more detail below, and it is the only known way that resolves the disquieting classic paradox of increasing entropy.

Chapter 9

Time and Antitime

"… sanza Dio non si volge in ramo foglia"
[*"… without God no leaf turns into a branch"*]
from *Rappresentazione di san Giovanni e Paolo* (1491)
Lorenzo di Piero de' Medici (1449–1492)

"Vita
La vita è fatta di ricordi, presente e pensieri,
il presente è già diventato passato e un ricordo,
il prossimo futuro è già diventato presente,
il lontano futuro è solo un pensiero.
La vita è fatta di ricordi, presente e pensieri."
[*"Life*
Life is made of memories, present and thoughts,
the present has already become past and a memory,
the near future has already become present,
the distant future is just a thought.
Life is made of memories, present and thoughts."]
24th May 2017

"Canta in un attimo il Suo eterno silenzio
Laddove non scorre nemmeno il Mio tempo:
In cima al Mio mondo, sul Colle del Vento"
[*"He sings in an instant His eternal silence*
Where not even My time passes:
On top of My world, on the Hill of the Wind"]
11th August 2019 (see dedication)
Francesco Spadoni (22nd July 2002 - 8/9th January 2024)

……………………………………

What is time? Why does time only ever flow in one direction? Is there a secret to the quest for immortality? Does a "*time machine*" exist?

These questions, by no means idle, have fascinated the most "active" minds of men for thousands of years, including, e.g., Saint Augustine, but also the most brilliant and famous theoretical physicists who constructed today's knowledge. It has been discussed for a long time and - concerning the more "serious" and professional aspects - many scientific studies have focused on the single - or cross - properties on how the known laws of theoretical physics are invariant (i.e., they do not change or change slightly) when considering the change in "parity" (P invariance, i.e., changing the formulas in a mirror image, just as the right hand is mirror image of the left hand), or the change in electrical charge (from positive to negative or vice versa, called C invariance), or by reversing the direction of the flow of time (called T invariance).

All of this reminds – ultimately – about a paradox. At the end of our journey, all of this will no longer be a paradox, and all these questions will find an answer that we believe is convincing.

The key rationale for our argument is based on Fig. 4.1, reproduced here again as Fig. 9.1. Contrary to what is generally taken for granted, the world of the Absolute is that of the timeless world, the world β. That is, unlike what is generally considered obvious, or almost innate, "time" does not have an absolute nature.

Indeed, in asserting this, one is not saying anything new at all. In fact, at the origin of the birth of human thought, a heated multi-millennial debate existed on the absolute value or not of "time". An explicit and highly credible testimony is found in *Genesis* and in the sacred Hindu texts, written - it is estimated - around 1500 BC. That is, we must enter into the realm of the archæology of the concept of "time", extensively discussed in a work dedicated to this topic, which it is not necessary to recall in detail here.

Those who conceived *Genesis*, or the Hindu sacred texts, believed that the Absolute world - the "Divine" one of the Absolute God - is timeless. In a world without time it is impossible to number anything, as to number you must first say what comes first, then what comes second, etc. Thus, in *Genesis* it is claimed that God is One and Triune. What does this mean? It means that God is uncountable because He is outside of "time". God can be One or made up of many entities, and the choice is only ours, totally arbitrary.

The Trinity of *Genesis* finds a parallel - although with substantial differences in attributes - in the Trimurti of the Hindu religion. The Trimurti (three forms, in Sanskrit) - often also referred to as the "Hindu Trinity" - indicates the three main divine aspects, manifested in the forms of three important archætypal Devas: Brahma is the Creator; Vishnu is the Maintainer; Shiva is the Destructor. It is significant that the origin of these concepts occurred almost "simultaneously", perhaps around 1500 BC in the Western world and the Indian one, denoting a possible or probable cultural contact between the two worlds.

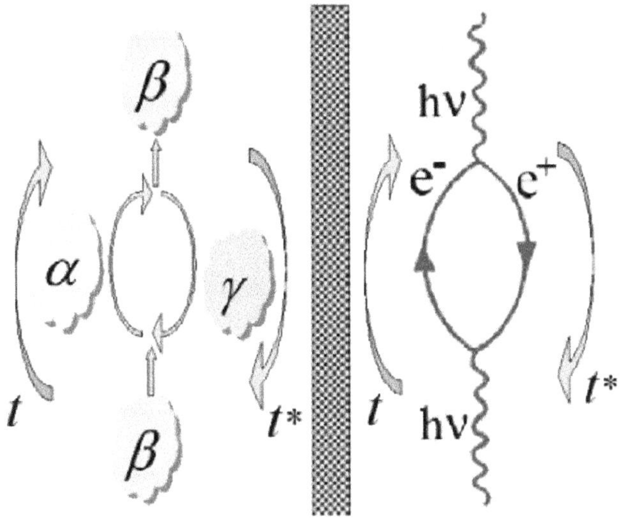

Fig. 9.1. In the right panel, a photon with energy $h\nu$ is indicated, which materializes in a pair of an electron (e^-) and a positron e^+ (more properly called antielectron). The phenomenon is paradoxical and unexplained, and is observed regularly, but the phenomenon also concerns other particles, other than electrons and antielectrons. This is a mystery that we must explain. All of this can be interpreted (left panel) with the existence of three worlds. The "absolute" world is the β world that is timeless. Occasionally, from the world β, objects are generated in two symmetrical and mirrored worlds α and γ. In world α time flows as we know it. In the γ world time flows in the opposite direction, from the future to the past, i.e., we have an antitime. This logical scheme seems to be the only possible one to solve the mystery of antimatter. The concept of a timeless "absolute" world is not new, but it is found in *Genesis* and in sacred Hindu texts.

In the classical world of ancient Greece - after a now forgotten heated debate - the concept of absolute "time" of the great Greek philosophers finally prevailed, which has been handed down to us and today is generally considered innate and well established.

In the Christian world, there is still an authoritative testimony of the great original debate in the philosopher/theologian Origen, who can easily be defined as the greatest scholar of Christian antiquity.

Origen was born around 185 AD, probably in Alexandria. Together with his family, he fell into great destitution in the years 201/202 following the death of Leonidas, his father, a Christian martyr of the persecution of Septimius Severus. At just 16-17 years old he tried to provide in some way for the basic needs of his large family by giving lessons. In 204, Bishop Demetrius placed him, despite his young age, at the head of the first official catechistic school. Origen immediately took his friend Heraclas as an assistant in the teaching of grammar, while he reserved for himself, for the more advanced students, that of philosophy, speculative theology, and Holy Scripture. In the early days, he also followed the lessons of the famous Neoplatonist Ammonius Sacca - also a teacher of Plotinus - whose method and philosophy were to assume great importance for Origen's theology. Origen also gained some knowledge of the Hebrew language. He led an ascetic life and in 210/211, falsely interpreting the passage of the sacred writings (Matthew, XIX, 12), he castrated himself.

Origen interrupted his teaching with frequent journeys. In Alexandria, he was helped by his friend Ambrose, whom he had diverted from Gnosticism. Perhaps following the massacres ordered in 215 in Alexandria by the emperor Caracalla, Origen left the city and went to Cæsarea in Palestine, where he held conferences to the community, at the request of bishop Theoktistus and also of Alexander, bishop of Elias Lapitolina (Jerusalem).

Fifteen years later (in 230), when traveling through Greece, he arrived in Cæsarea, and his two bishop friends ordained him a priest, despite his voluntary mutilation. Angered by this fact (emasculated people could not be ordained), his bishop, Demetrius had him declared deposed from the chair and from the priestly office by two Synods held in Alexandria (230/231), on the grounds of the irregularity of his ordination and probably also for some opinion not admitted by the Church.

Origen then went to Cæsarea where he remained until the empire of Decius, and where he founded a school similar to that of Alexandria. His disciple was Gregory the Thaumaturge.

His scientific production had made him famous even among the pagans. Iulia Avita Mamæa, mother of the emperor Alexander Severus, invited him to Antioch between 218 and 222 to hear his conferences. Under Decius he was probably imprisoned in Cæsarea itself and barbarously tortured. He died in Tyre, as a consequence of this treatment, at the age of 70, probably in 253.

Origen can be considered the first great Christian thinker who competed with the dominant philosophical thought of the Greek classics. Origen was really obsessed with explaining the mystery of the Divine Trinity, and from his vast literary production (e.g., by carefully reading *De Principiis*) we can see how he was very close to the correct concept of non-absolute "time". However, he was too sharp and advanced a thinker compared to his contemporaries - who were therefore often inclined to consider him a heretic, as his adventures and troubles show. For this reason, Origen often backed away from the final formulation.

Now it seems paradoxical that an ancient concept of "time", which dates back to around 1500 BC, must necessarily be rediscovered to get to grips with the paradoxes of the most advanced modern physics!

Thus it happened that Origen, since his time, was considered the most important theologian of the Greek Church. His name was discussed in Christian antiquity and uttered with enthusiasm or contempt. If many heretics invoked his authority, true doctrine also had much to learn from him. Origen wanted to be an "orthodox" Christian, which is evident first of all from the enormous value he attributes to the official teaching of the Church, so much so that he considers every error of doctrine worse than a moral deviation. However, his preferences for the allegorical interpretation of *Scripture* - and the influence of Platonic philosophy - made him fall into what were then considered serious dogmatic errors. The discussions surrounding his orthodoxy ended only three centuries later, in the 6th century, when the emperor Justinian I, in the edict of 543, condemned nine of Origen's propositions. All the bishops of the Empire soon adhered to this edict, including the Patriarch of Constantinople, and Pope Vigilius, among the first. Origen's brilliant thought thus "officially" entered the limbo of curiosities to be forgotten.

Origen surpassed in literary fruitfulness all the Fathers of Christian antiquity. Many of his works are occasional productions, such as sermons and conferences, which were recorded by stenographers, and only in this way can we understand the enormous number of writings, and their not great literary value. Of his numerous works - mostly of a biblical nature - only a small part has reached us, and not in the original Greek.

Incidentally, it is reasonable to hypothesize that the deceased enter the β world - and in this regard, see the short poem in the dedication of this volume, and in the title of this chapter, written by an adolescent poet, who had earned from his grandfather some very simple concepts, and which he was able to summarize incredibly in just a few words. The poet was our beloved grandson and nephew, who died immaturely and unexpectedly, leaving us all inconsolably heartbroken.

As for the centuries to follow, we cannot report in a few lines a more than lively and intricate philosophical and theological debate. The debate in the Christian world on the meaning of "time" survives very intensely even to the present day, but it is found in the sphere of the great ascetics and partly in the philosophical sphere, while the scientific - and more general historical/philosophical - sphere the "dogma" of "absolute time" has now been passively accepted as we mistakenly conceive it today in the natural sciences, almost as a natural and innate concept.

But "time" is *not* absolute, it has value only in the world α, the world of matter and entropy (see Chapter 8) - while similarly and symmetrically in the world γ the concept of "antitime", of antimatter, applies to the antientropy.

There are (see Chapter 4) galaxies, or clusters of galaxies, composed of matter, and galaxies or clusters of galaxies made up of antimatter. When these galaxies and antigalaxies "annihilate" and enter the β world, entropy and antientropy cancel each other out. New galaxies and antigalaxies of matter and antimatter will be born by "materialization" in a mirror-like manner in the α and γ worlds, and the cycle will repeat itself outside of time, in the timeless β world.

Let's make a short, slightly specialized excursion - for the reader who has a little knowledge of mathematical analysis. The worlds α and γ are "*conjugate complex*" of each other. That is, in the past, it was said that Nature expresses itself through mathematics derived from the succession of *integer numbers,* through *relative numbers,* through *rational* and *irrational numbers,* etc. The notion of "*complex numbers*", i.e., made up of a "*real*" part and of an "*imaginary*" part, has always been considered a brilliant tool invented by mathematicians, suitable for defining some abstract algorithms, which are useful for logic and calculation, but divorced from the reality observed.

Conversely, Nature must be considered as intrinsically represented by "*complex numbers*", and the two "*conjugated complex*" faces are the expression of the existence of two mirrored and parallel worlds, the world α and the world γ, of matter and antimatter, of time and antitime, of entropy and antientropy, etc. In other words, this is the dualism represented by the d'Alembert equation.

We will see this better when we come to apply a "statistical" treatment to the atom monad with the Schrödinger equation. This equation and "statistic" allows us to measure only the amplitude of the continuous function Ψ which describes the atom monad observed from the outside. In contrast, it is not possible to measure the phase of the continuous function Ψ. In other words, this limitation is also a consequence of the empirical constraint due to the fact that we can observe the atom monad only from outside.

But let us now come to the concept of the mysterious one-direction of time - and how this is connected to the search for immortality, or to the imaginative "*time machine*".

Let's remember Chapter 7, where we considered the reasonable hypothesis that every natural system is constantly evolving in the search for equilibrium without ever reaching it. That is, every natural system - including the entire Universe - evolves according to states of quasi-equilibrium, which progressively approach the final state. However, nothing ever reaches final equilibrium – and, incidentally, we hypothesized that, perhaps, appropriate subsets of the sequence of Fibonacci numbers represent different “paths”, characterizing different sub-systems, in the search for the final ideal equilibrium.

We also considered that if Nature proceeds through states of equilibrium that are actually achieved, the entire creation would be a desolate swamp, totally dead with no evolution, without life, without perception of “time”. The whole scenario would be a very bleak scenario.

The “perception” of time is, in fact, linked to the observation of something that evolves - and everything evolves, and without evolution there would be no life, but neither would there be death. Every natural system, even stars or galaxies, or galactic clusters, have a life cycle, they are born and die ... that is, Nature in itself is intrinsically irreversible, it always evolves with “time” (or antitime)!

Thus *immortality* cannot exist in any natural system. But Nature also knows how to overcome this terrible inconvenience, and the key comes from the so-called “*antifragility*” that we will discuss in Chapter 26. That is, Nature has foreseen that every natural system - even biological - can regenerate itself through reproduction. That is, single individuals and/or single natural subsystems die – with no exception - but not before having generated new young individuals, and in this way the “transience” of “time” is overcome in the sense that every natural system obviates its death by permanently reproducing younger identical or analogous systems (autopoiesis).

We will return to these fundamental concepts in greater detail below (in Chapter 26, in Chapter 27, and in Chapter 28). But we now want to focus on the mysterious and much debated invariance with respect to the reversal of the direction of time, which is notoriously observed in every known law in physics - and has been and still is the subject of careful studies and debates even in the most advanced theoretical physics.

In fact, all laws have always been indiscriminately - consciously or unconsciously - conceived in accordance with the “original sin” or logical “bug” of Newton, who hypothesized a direct, immediate connection between cause and effect. That is, every action at a distance has always been conceived *ab origine* as instantaneous propagation at infinite speed of the signal. By definition, states of strict equilibrium determined by precise chains of cause and effect have always been considered. All natural science, physics, chemistry, biology, medicine, engineering, etc., is based on this *a priori* assumption, generally unconscious. This conception survives in the study of gravitation - and only in electromagnetism there is the exception of the propagation of the Poynting vector at the speed of light, and this has led to paradoxes that gave rise to relativity (as we will see in Chapter 16) - and, as shown in Chapter 6, led to the concept of monads.

In other words - in compliance with the wise Galilean criterion - every experimenter tries to isolate different individual phenomena and he studies how different conditions can influence every given phenomenon.

Thus, we look for "laws" that regulate every known phenomenon in quantitative terms. They are first and foremost empirically verified laws, trying to minimize the difficulties imposed by the empirical constraint. All of these "laws" are blessed – and are the backbone of all engineering. The "monument" of mathematics is the "language" necessary to best exploit all this knowledge. Then we try to frame these empirical "laws" in a theoretical, axiomatic scheme, and this is the task of theoretical physics… The whole is a wonderful construction of the human intellect!

However, if by definition - both consciously and unconsciously - all the known "laws" have been imagined as relations between *states of equilibrium*, how is it possible that they can reveal any failure and do not strictly respect the invariance with respect to the inversion of the direction of time?

In fact, the perception of "time" is intrinsically and inextricably associated with *irreversibility*, i.e., with the evolution of every natural system according to states of quasi-equilibrium, while the final equilibrium can never be achieved. If the final equilibrium were reached, the entire Universe would become a totally dead swamp, and paradoxically, the Universe would die completely and also quickly.

When we limit our attention to a partial natural system, to comply with the Galilean recommendation, we inevitably lose the perception of irreversibility, and the "laws" we derive are necessarily *reversible* over time. On the other hand, Galileo's teaching was blessed, as it was decisive for the development of engineering.

Therefore, fortunately, "time" flows in only one direction – i.e., Mother Nature has the characteristic of *irreversibility*. It is fortunate that there is the death of natural systems, because *without death there would be no life*!

Finally, let us also remember an ethical aspect that has been - and still is - the subject of extensive debate, and is linked to the dictum "*no leaf moves unless God wants it*". The objection is that this rule would deny the free determination of every individual to choose the path of good or that of evil etc. Free choice lies in the fact that - among multiple states of quasi-equilibrium during the evolution of his life - the individual has the faculty to direct his life within certain limits towards objectives that seem more or less in conformity with his ethical and existential choices.

That is, there is always a large, although limited, margin of decision left to the arbitrariness of the individual. However, every individual is a part of the "Universe" system, and his "evolution" is inevitably conditioned by the evolution of the Universe. The fate of every sub-system of the Universe, and of every individual, can be nothing other than death.

Having exhausted the discussion on the concept of "time" and "antitime", let us now delve into the structure of the photon, and of particles and antiparticles, in greater detail.

Chapter 10
Photons and "*Miniphotons*" – Volleys of "*Miniphotons*" – "*Elements*" and Waves – Spirals and Waves as "*Statistical*" Models – Chirality – Stability of a Photon/Antiphoton

"The important facts are the crucial facts ... that is to say, those which can confirm or invalidate a theory. After this, if the results are not in accord with what was anticipated, real scientists do not feel embarrassment which they hasten to eliminate with the magic of handwaving; on the contrary, they feel their curiosity vividly excited; they know that their efforts, their momentary discomfiture, will be repaid a hundredfold, because truth is there somewhere, nearby, still hidden and, so to speak, adorned by the attraction of the mystery, but on the point of being unveiled."

Jules-Henri Poincaré (1854-1912)
[*Dernières pensées*, 1913]

....

The "game" now becomes a little "heavier". We must consider an unusual aspect of the internal structure of a photon, associated with a fundamental property in Nature, linked to spirals which - as we will see - are a property shared by all natural systems, from the subatomic world up to galaxies or galactic clusters.

"Playing" with the spirals - and with their dual nature depending on whether they wind clockwise or anticlockwise - requires a little care in observing with particular attention some specific details of graphic representations. The concepts are simple, but must be tackled with concentration. The reader should not feel discouraged. It's just a matter of playing with a kind of abacus, where, however, every ball that moves can have a profound physical meaning. It's just a patience problem.

Photons and microphotons

Let's consider the two electromagnetic waves - with specular parity - of Fig. 6.2. We take into account the error due to the continuity hypothesis, and we separately plot the two specular waves that correspond to the two solutions of the d'Alembert equation (see Chapter 6), where however each "wave" is transformed into discrete "*elements*", which we will call κ_E and κ_H. These two types of "*elements*" correspond to the electric and magnetic field component, respectively.

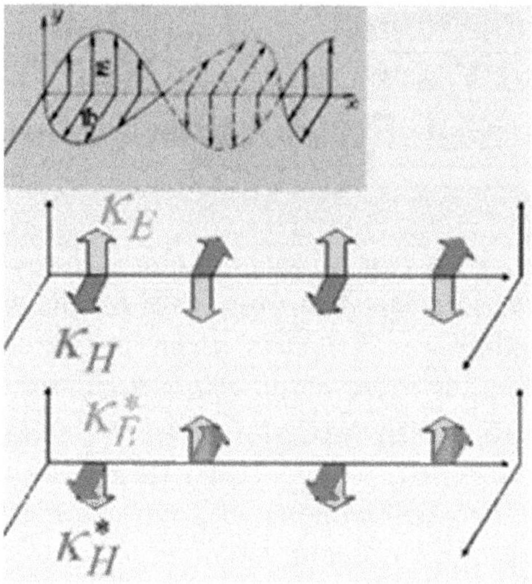

Fig. 10.1. [*Upper figure*] Standard representation - as in electromagnetism treatises - of an electromagnetic wave polarized in a plane. [*Central figure*] the same concept shows schematically a segment of a "*volley*" of "*miniphotons*". Every "*miniphoton*" is composed of two pairs of "*elements*" perpendicular to each other, i.e., each component of the pair contains an electric "*element*" κ_E and a magnetic "*element*" κ_H. [*Bottom figure*] The same concept with parity reversed. The asterisk in the lower graph is redundant, since the "*elements*" are the same as in the central graph, while only the parity of the wave, i.e., the relative orientation of the "*elements*", changes. The origin is on the left – and the origin is the same for the three graphs – although the wave proceeds to the right according to the time in the upper and central graphs – and according to the antitime in the lower one. However, antitime flows in the direction opposite to time. See text.

The continuous wave represents an interpolation, by means of continuous functions, which is arbitrarily performed on a phenomenon which in reality is quantized. As a curiosity, in Chapter 6 the two spirals were assimilated to the staircase of the *Vatican Museums* or that of the *St. Patrick's Well* in Orvieto. The miniphotons are the equivalent of the steps, which in turn have a structure with a vertical part and a horizontal plane, etc.

Every photon is made up of a "burst" - which we prefer to call a "*volley*" - of "*miniphotons*". Every "*miniphoton*" has an internal structure, being composed, respectively, of a pair of two "*elements*" κ_E and κ_H, i.e., every "miniphoton" in Fig. 10.1 contains two "*elements*" κ_E and two "elements" κ_H.

Let us now superimpose the two middle and lower figures of Fig. 10.1, and represent the result as in Fig. 10.2.

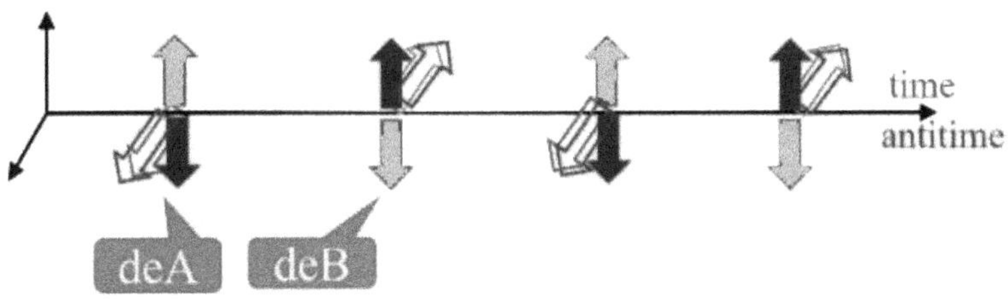

Fig. 10.2. The second and third graphs of Fig. 10.1 are overlaid. The two "*elements*" κ_E and κ_H are represented by blue (or light grey) arrows - respectively, full color or only with outline. In the case of a photon of standard polarity, i.e., in the case of the blue (or light grey) helix in Fig. 10.4, the "*elements*" are related to the "time" axis. Symmetrically, the two "elements" κ_E and κ_H, which refer to the photon of opposite polarity (i.e., the antiphoton), are indicated with the red /dark grey) color and are referred to the "antitime" axis. However, let's consider that there is no distinction between photon and antiphoton, since they must all coexist together within a single entity, which for brevity's sake, we still call "photon". If photon and antiphoton are separated, this is the "materialization" of a photon into matter and antimatter. See text.

We must therefore distinguish - with patience - two types of "*miniphotons*", which denote a different parity depending on the mutual orientation of the "elements" κ_E and κ_H. In the α world, standard photons move at the speed of light c referred to as "time". The photons of opposite parity – i.e., the "*antiphotons*" – are observed only in the γ world and move at the speed of light but referred to the "antitime", i.e., to a direction reversed with respect to "time".

For simplicity, we will call a "photon" the set of a "photon" and an "antiphoton". A "photon" is in fact a combination of two "*volleys*" of "miniphotons" of identical length and opposite parity - which, in any case, are always closely linked to each other. One of the two "*volleys*" propagates through time, the other through antitime. Therefore, the photon cannot have its own intrinsic "time" - given that it contains a component referring to time and one to antitime - i.e., the photon is part of the timeless world β.

In a world without time, no form of speed can be defined. Therefore, a photon in the world β is immobile, while a photon that is observed in the world α moves with speed c, and a photon that is observed in the world γ moves with speed c, but referred to the antitime which flows in the opposite direction to time.

Note that this reasoning applies to any frequency, even to radio waves - as long as it is possible to apply the continuous interpolation approximation implicit in the concept of "wave".

It should be stressed that the representations shown in Fig. 10.1 - both in the upper Fig. 10.1 and in the middle and lower ones - are only formal, conventional images, which we have arbitrarily chosen as purely mnemonic graphic symbols. In reality, due to the empirical constraint, we can only observe the "photon" monad from outside. Therefore, the true, internal structure of the "photon" is unknown to us due to the empirical constraint. In other words, we can only know that, by interpolating the quantized structure that exists in nature, we observe electromagnetic waves. Any other more detailed model is inevitably the result of our imagination and of the formalism that - for mnemonic simplicity - we choose to use.

Therefore, in what follows, we will prefer to change Fig. 10.2 to Fig. 10.3. In Fig. 10.2 and 10.3 we have superimposed the middle and bottom Fig. 10.1 in a single "*volley*". But Figs 10.1, 10.2 and 10.3 have nothing physically different from one another, other than the choice of the conventional graphic symbol for each "*element*" which is part of every pair that makes up a "mini-photon".

We now introduce the further hypothesis that every pair of "*elements*" κ_E and κ_H must be located on the same abscissa (time or antitime) in the diagrams of Fig. 10.3. Thus, Fig. 10.3 shows two components, each one being the superposition of two pairs of "*elements*" κ_E and κ_H. We call every such component a "*d-element*" with two pairs of "elements". Fig. 10.3 is physically the same object represented by Fig. 6.4, reproduced here as Fig. 10.4.

Note that the blue (or light grey) helix of Fig. 10.4 (and respectively the red or dark grey helix), is the continuous interpolation over the blue-light grey (red/dark grey) "elements" of Fig. 10.3. Furthermore, as mentioned in Chapter 6, with an elementary formal mathematical transformation (which is found in every discussion of electromagnetic waves and which there is no need to report in detail here) it is always possible to decompose an electromagnetic wave into two polarized components. This decomposition can equally be carried out in perpendicular planes (as above). Or, equivalently, the decomposition can also be performed in circularly polarized waves – i.e., in two circularly polarized components, where one component is a wave with a spiral in the clockwise direction, and the other a wave with a spiral in the counter-clockwise direction.

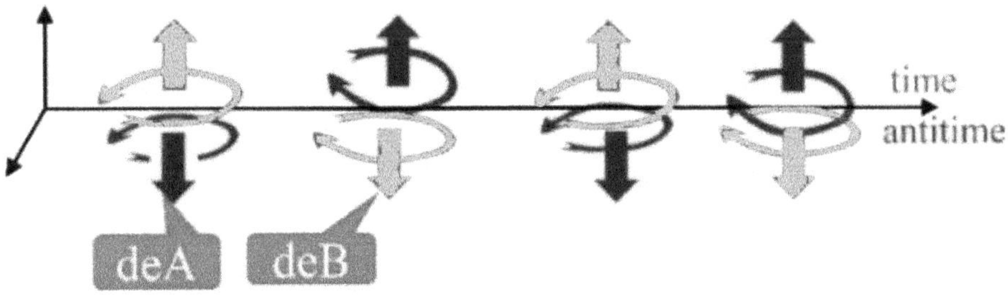

Fig. 10.3. Alternative representation of the same concept expressed in Fig. 10.2, through a different conventional graphic "language". That is, instead of empty arrows, we choose two symbols with a spiral - which better denotes the mirroring of the "*elements*" κ_E and κ_H. Particles are conventionally associated with blue (or light grey) symbols and antiparticles with red (or dark grey) symbols (see Chapter 12 and Chapter 13). The association of the blue (or light grey) arrows with anticlockwise rotation is arbitrary. It is important, however, that the direction of rotation is opposite between the α world and the γ world. The symbols *deA* and *deB* denote two types of coupling between "*elements*", as will be better seen below. The symbol "*de*" stands for "*d-element*", or "*double-element*", and every "*de*" can be of two types, A and B, precisely. See text.

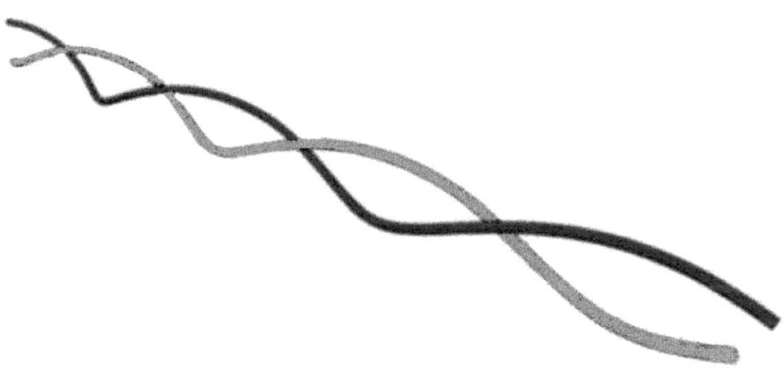

Fig. 10.4. If the two specular electromagnetic waves are superimposed - which correspond to the two solutions of the d'Alembert equation, transformed by the polarization according to orthogonal planes into circular polarization (see text) - you have a structure with two linked spirals. One spiral travels in one direction (that of "time", here in blue or light grey), the other spiral in the opposite direction (that of antitime, here in red or dark grey). This structure is typical of a "photon" which - having a component that flows with time, and a component that flows with antitime - is outside of time and antitime, i.e., it is in the β world. This double spiral structure recalls the famous "*St. Patrick's Well*" in Orvieto, or the visitor access/exit staircase at the *Vatican Museums* in Rome.

Therefore, the separation into circularly polarized components better expresses the transition from the conventional representation of Fig. 10.2 or 10.3 into that – interpolated and averaged with continuous functions – shown in Fig. 10.4.

The concept of spiral structure, clockwise or anticlockwise - of any natural system - is commonly referred to as "*chirality*" (see Chapter 26).

The stability of a photon/antiphoton double helix

Consider Fig. 10.2, 10.3 or 10.4. We ask ourselves: how can a double helix survive - i.e., how can the double helix maintain its internal structure and avoid splitting into two helices? That is, how can a photon avoid "materializing" in matter (which is part of the α world) and antimatter (which is part of the γ world)?

As already stressed, the meaning of the wave in the upper graph of Fig. 10.1 must be considered. A continuous wave is only an interpolation, carried out by a continuous function, that superimposes the real physical phenomenon, which is certainly discrete and quantized. Indeed, the real physical phenomenon is more correctly represented by the graphs of the second and third rows of Fig. 10.1. The same applies to Fig. 10.2 or 10.3, transformed into Fig. 10.4, which is a superposition of two continuous 3D functions averaging over structures quantized into discrete components. The discrete components are miniphotons, and every miniphoton is composed of two pairs of "*elements*" κ_E e κ_H.

Let's anticipate a concept more fully discussed in Chapter 12, and suppose that an "element" κ_E pointing upwards (in our arbitrary conventional graphical representation) is associated with a positive electric charge - and that an "element" κ_E pointing downwards is associated with a negative charge. But note that there is no physical reason or implication in this assumption, as it is merely a conventional symbolic choice, and no physical meaning is associated with "up" and "down."

By the way, a "technical" note. For the moment, we do not consider the spin and the intrinsic magnetic moment of particles and antiparticles, which are physical entities defined at a later time (see Chapter 12).

That is, the simple association with the charge of "elements" κ_E pointing upwards or downwards is just an arbitrary choice of axiomatic symbolic representation of "electric charge" – the existence of which is an unambiguous fact proven by rubbing electricity (see Chapter 5) - without, however, that "up" and "down" have any physical implications ("up" and "down" compared to what?). The spin and magnetic moment are independent physical entities that are not related to charge. Their axiomatic symbolic definition is considered below in Chapter 12, regardless of the axiomatic symbolic representation of the charge.

Since the two κ_E *"elements"* in every Fig. 10.2, 10.3, or 10.4 are oppositely charged – and one *"element"* participates in the photon helix, and the other *"element"* participates in the antiphoton helix – it is reasonable to assume that the two *"elements"* κ_E attract each other and give stability to the double helix. That is, the double spiral structure is a sort of stable configuration, and the attraction between κ_E *"elements"* is the *"glue"* that prevents the doubling, i.e., the "materialization" of the double helix into a particle and an antiparticle.

Furthermore, it is also likely or very probable that an even stronger *"glue"* is provided by the two overlapping κ_H *"elements"*, having a presumably intense mutual coupling. Note that, in this argument, the belonging to photon or antiphoton of both the *"elements"* κ_E and κ_H is assumed to be indistinguishable, whether an *"element"* is part of a *volley* of miniphotons of a parity, or whether it is part of one *volley* of the other polarity.

In any case, it must be again emphasized that the association with graphic symbols is only the arbitrary choice of a "language" that can be modified in the future whenever needed, in order to correspond to the growing amount of observational details, and compatibly with the empirical bond.

In conclusion, the photon has a double spiral structure, more specifically made up of discrete objects, or miniphotons.

The two spirals are characterized by opposite parity or polarity, i.e., the two spirals have opposite directions (one winds clockwise, the other counterclockwise).

The two spirals are linked to each other thanks to the *"glue"* represented by the two κ_E *"elements"* associated with the electric charge, and by the two κ_H *"elements"* which have magnetic properties. When these *"glues"* are missing, the two overlapping spirals decouple, and one spiral enters the α world, the other the γ world.

The two spirals - when they are linked in the photon - are part of the β world, which is timeless, therefore without the possibility of defining the speed. Hence, the photon in the β world cannot have any velocity. When the photon is observed in the α world or in the γ world, the photon is only seen moving with the speed of light c, related in the two worlds to time or antitime, respectively. After decoupling, the spiral in world α moves at *a speed always lower than the speed of light* c, and symmetrically, the same happens for the spiral in world γ.

In addition, each of the two decoupled spirals, when each spiral is observed in the respective world α or γ, has a content of emp or antiemp (i.e., negative emp). The photon has a content of emp or antiemp, depending on whether the photon is observed in the α world or in the γ world.

However, each emp or antiemp in the world α or in the world γ, respectively, can always move only with a speed lower than the speed of light c in the respective space. That is, we can say that when the double spiral "materializes" following the decoupling of the two superimposed spirals, the emp and the antiemp "*degrade*" given that they can no longer move at the speed of light c.

It all certainly seems imaginative. After all, Newton's definition of "force" is no less "imaginative". We accept the definition of "force" because it works very well but only for engineering applications. However, we must consider it an averaged concept, suitable for the description of macroscopic phenomena of everyday life. However, when we enter the atomic and subatomic world, new axioms are required. This is the task of theoretical physics which must certainly reject every logical structure that leads to paradoxes.

In any case, the logical scheme proposed here appears to be the only known possible explanation that allows us to overcome painful paradoxes that have plagued theoretical physics for well over a century. Any alternative proposal of the formulation described here is always welcome, as long as it is capable of overcoming the paradoxes currently existing according to the conventional treatment.

Chapter 11

Doppler Effect – Energy of a Photon

"We do not believe any group of men adequate enough or wise enough to operate without scrutiny or without criticism. We know that the only way to avoid error is to detect it, that the only way to detect it is to be free to inquire. We know that in secrecy error undetected will flourish and subvert."

Julius Robert Oppenheimer (1904-1967)

....

Let's now consider a classic topic of physics, i.e. the Doppler Effect. This is the apparent change - compared to the original value - of the frequency ν of a wave, as it is perceived by an observer when the wave source is moving relative to the observer. For instance, Fig. 11.1 shows the case of sound waves emitted from a source moving towards the left.

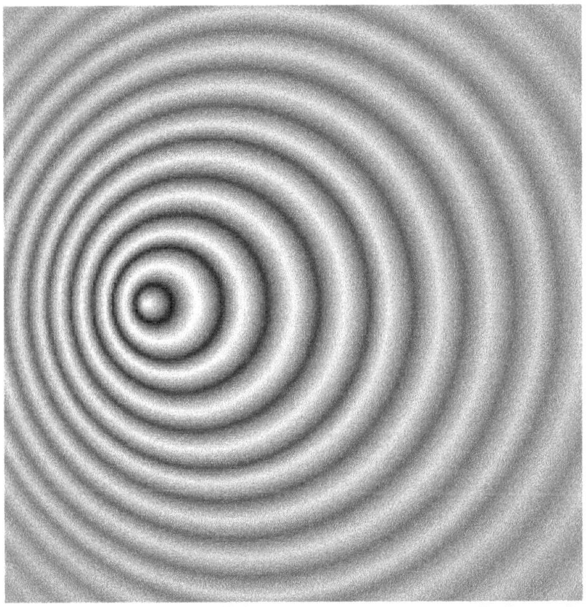

Fig. 11.1. Sound waves emitted by a source moving leftward. For example a vehicle (car or motorcycle or ambulance) moving at high speed on a straight road emits a sound that the human ear - as is well known - perceives differently during the approach and moving away of the vehicle. In the case of a plane or rocket, which moves in the air at about 1,000 km/h, all the spheres shown in this figure touch each other on the left, and the so-called "wall of sound" is formed which subjects the cockpit of the aircraft to very great efforts, even resulting in the destruction of the aircraft.

The Doppler Effect was originally studied in 1845 by Christian Andreas Doppler (1803-1853), an Austrian mathematician and physicist. In 1848, the French physicist Armand Hippolyte Louis Fizeau (1819-1896) independently discovered the same effect in electromagnetic waves. The Doppler Effect has numerous applications in both astronomy and medicine (for the study of blood flow), etc. Let's consider how the Doppler Effect should be interpreted according to the structure of the photon discussed in Chapter 10.

As a premise, let us remember that we must assume that an "absolute" reference system certainly exists in Nature, which we call "æther" (see Chapter 4). In fact, it could also be that we cannot identify the æther with respect to our reference frame. However, for our discussion, we must admit that the "absolute" system "æther" exists, even if we may eventually fail to recognize it.

Let's consider Fig. 11.2. A "classic" problem in physics is represented by the energy of a photon. As is well known, in all physics texts, it is stated that a photon has an energy $h\nu$ associated with it, where ν is the frequency of the light associated with that photon, and h is a constant, called Planck's constant.

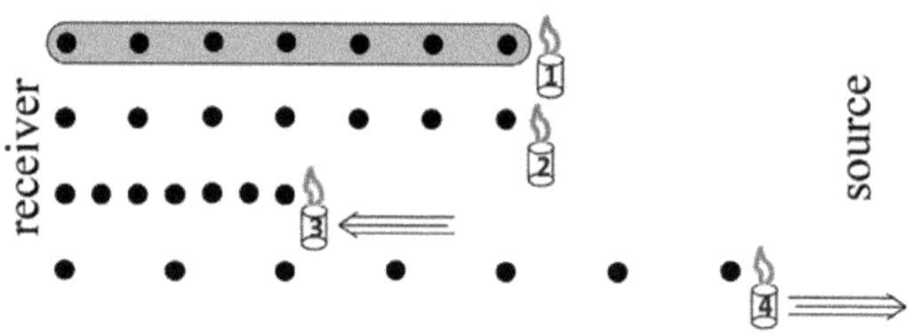

Fig. 11.2. The Doppler effect. A photon propagates to the right either like a unique bullet or like a volley of miniphotons. Every miniphoton is represented here with a black ball. *(1):* A photon is represented here as a single "bullet" (i.e., a single quantum) (gray patch). *(2):* The photon is represented by a volley of approximately equidistant miniphotons, released at approximately constant time intervals. The source of light is stationary. *(3):* The same as case *(2)* when the source moves leftward, i.e., in the direction opposite to the propagation of light. *(4):* The same as case *(2)* when the source is moving in the same direction as light. Miniphotons always move at the constant speed c of light, referred to the "absolute" reference system, i.e., to æther.

As far as our logical journey is concerned we must be more cautious.

Let's consider first of all that we have defined the concepts of emp and antiemp. But we still haven't defined the concept of "energy". The concept of kinetic energy is defined only in Newtonian mechanics which we wanted not to consider for now. For Newton, a particle of mass m and speed $\mathbf{v}$ has kinetic energy $(1/2)m\,\mathrm{v}^2$ (see Chapter 4). However, in the case of the photon, it is always $m = 0$, and $\mathbf{v} = c$ is constant. Therefore, for our discussion, it is much more appropriate to correctly consider the quantity of emp associated with a photon or miniphoton as it appears in an interaction between a photon and a particle or antiparticle (see Chapter 14, Chapter 15, Chapter 18 and Chapter 21). In today's physics, the energy of a photon results from all the available experimental measurements, studying the emission "*spectrum*" of various substances.

We explain what the "*spectrum*" is. When a substance is heated to the point of emitting light, we study how many photons are emitted for every given frequency ν. The result is called the "*spectrum*" of that substance, and the spectrum is interpreted with the Schrödinger equation which describes - as far as possible - what happens inside the atom monad that emits the photons and is observed from the outside.

Therefore, every photon that we can observed and measure is generated from a monad atom. The atom monad is observed from the outside, and is governed by the Schrödinger equation, while the emission of a photon always occurs through the "slingshot effect" (see Chapter 6). That is, we can only - and always only - experiment by means of photons which – in Fig. 11.2 – are represented by the "bullet" (quantum) shown as a single object as a single gray patch (in the top plot of Fig. 11.2). In any case, every photon thus generated contains a quantity of emp which - in the α world - is correctly expressed as $h\nu$, *in agreement with the requirements of the Schrödinger equation.* In other words, this result is a pure matter of fact of observation, and as such it is indisputable.

Now, if we change the observed frequency of the photon, as it changed due to the Doppler Effect - as done in the third and fourth rows of Fig. 11.2 - ν changes but the total emp content of a photon obviously remains the same. In other words, we must refer to the emp content associated with a single wavelength, which, in accordance with Figs. 10.1, 10.2 and 10.3, corresponds to a single pair *deA* and *deB* in Fig. 10.3. This will lead us to calculate the emp content in a single miniphoton.

To this end, we recall another concept well known in elementary physics – as shown by various optics experiments, often performed even in secondary school laboratories. We will limit ourselves here to describing only the inspiring principle. We talk about *interference* between two electromagnetic waves. Fig. 11.3 shows the case of constructive (upper plot) and destructive (lower plot) interference, and Fig. 11.4 shows a possible instrumental set used to observe and photograph the interference "fringes".

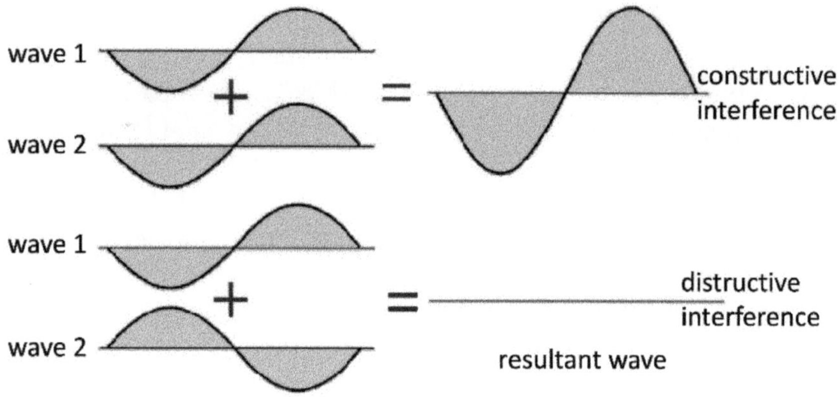

Fig. 11.3. The amplitudes A_1 and A_2 of two waves add together. If they are in phase with each other, there is constructive interference (upper plot). If they are out of phase by half a wavelength, destructive interference occurs. If the mutual phase shift is different from the two typical cases shown here, the interference gives an intermediate result between the two extreme cases.

The interference is constructive, or destructive, or intermediate, depending on the difference in the path traveled by the two light rays - this path for each light ray is called "*optical path*".

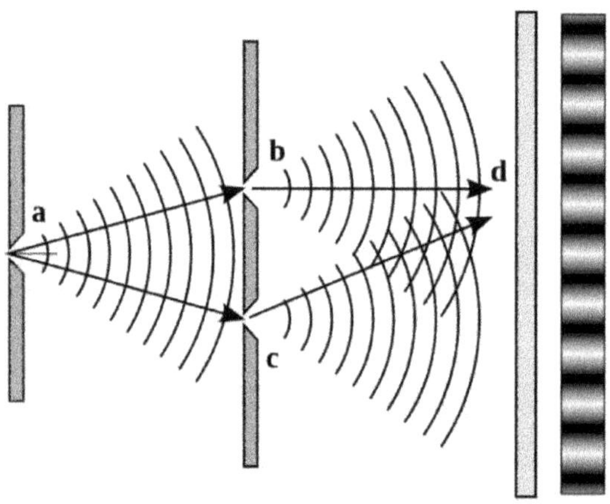

Fig. 11.4. A beam of light passes through a small gap a, and hits a second opaque barrier with two gaps b and c. The light that passes through the two slits b and c is equivalent to two distinct coherent sources of light (i.e., generated by the same source a), which, through constructive or destructive, or intermediate interference, impress a photographic plate d. This is one of many optical experiments, often as fun as they are delicate to carry out.

Further experimental data is very important. In fact, in principle the phenomenon should always be observed. In reality, when the difference between the optical path of the two light rays exceeds a certain value, the phenomenon is no longer observed. That is, it is as if the wave associated with a light ray has a limited duration. We then say that the wave propagates like a "*wave train*" and the "*wave train*" always has a limited length.

This says that the representation of the photon in the first row of Fig. 11.2 by means of a gray spot is realistic, since – as an unequivocal observational fact - the gray spot certainly has a finite extension. It is possible to measure this maximum difference in optical path of the two interfering rays of light - that is, it is possible to measure the length of the "*wave train*", which we will therefore call $L(\nu)$.

The explanation for the phenomenon is very simple. Every photon - ejected from an atom monad - can be approximately described with a wave, but this applies to a limited temporal extension. That is, it is said that it is necessary to consider the "*wave train*" of the observed wave, which is intrinsically limited. In other words, with reference to Fig. 11.2, the photon has a similar structure to the gray stripe of limited length in the first row. That is, a photon always has a limited number of miniphotons. This is a physical fact, incontrovertible.

From Figs. 10.2 and 10.3, we know that a wavelength λ corresponds to a couple *deA*+*deB*. Therefore, a photon is composed of a number of pairs *deA*+*deB* equal to $L(\upsilon)/\lambda$. If, consistently with the experimental evidence of atomic spectra and the Schrödinger equation, $h\nu$ is the total emp content of a photon observed in the α world, then the emp content ε of a miniphoton is (i.e., of a pair *deA*+*deB*) is equal to the total content of emp, i.e., $h\nu$, divided by the number of pairs *deA*+*deB* or divided by $L(\upsilon)/\lambda$. Furthermore, take into account the fact that $\lambda = c/\nu$, i.e., the wavelength λ is equal to the propagation speed c of the electromagnetic wave divided by the frequency ν, or in short

$$\varepsilon(\upsilon) = \frac{h\nu\lambda}{L(\upsilon)} = \frac{h\nu c}{L(\upsilon)\upsilon} = \frac{hc}{L(\upsilon)} \quad (11.1)$$

which in fact takes into account the fact that the "*elements*" κ_E and κ_H related to waves with different λ have a different extension in time or antitime (a light wave cannot be assimilated to a radio frequency wave).

In other words, the content of emp in every miniphoton is different for every given frequency υ. Therefore, just as the Schrődinger equation expresses in a "statistical" way the emp content of the photons ejected by the atom monad by "*slingshot effect*", an additional "*quantum number*" that characterizes every photon of a given frequency υ is the emp content of every one of its miniphotons $\varepsilon(\upsilon)$ which should be suitably investigated.

The result thus expressed refers to the world α, but it also holds in the world γ, where it must be considered that both antitime and the content of emp are negative.

Having thus clarified the internal structure of the photon as a volley of miniphotons, let's now consider how all this has consequences for the internal structure of particles and antiparticles, when we move from the β world to the α and γ worlds, which are characterized by "*degraded*" emp, that is emp which can only move at speeds less than the speed of light.

Chapter 12

Particles and Antiparticles – *QED* and *QCD*

"When a revision or a transformation of a physical theory is produced, one finds that at the starting point there is almost always a realization of one or several facts which cannot be integrated into the framework of theory in its current form. Facts remain always the key to the vault, upon which depends the stability of every theory, no matter how important. For a theoretician really deserving of the name, there is accordingly nothing more interesting than a fact which contradicts a theory which has been previously considered to be true, and thus real work starts at this point."

Max Planck (1858-1947) [*Initiation to Physics*, 1941]

....

In Chapter 10 we clarified the quantum and spiral structure of a *photon*, which should be conceived as a volley of miniphotons. When the discrete structure of the photon is interpolated with a continuous function, making a sort of smoothed representation, a double spiral is found – i.e., one spiral wraps in time and the other in antitime. The amount of emp of a miniphoton (observed in the α world or in the γ world) is

$$\varepsilon(v) = \frac{h\nu\lambda}{L(v)} = \frac{h\nu c}{L(v)v} = \frac{hc}{L(v)} \tag{12.1}$$

where h is Planck's constant, ν is the frequency of the photon, λ its wavelength, and c is the speed of light, and in fact $\varepsilon(\nu)$ depends on the observed wavelength, similarly to the distance between the black dots in Fig. 11.2. The result (12.1) refers to the world α, but it also holds in the world γ, where it should be considered that both antitime and emp content are negative.

Let us now consider particles and antiparticles - and their quantum and spiral structure, as happens when we pass from the β world to the α or γ worlds. Let us consider the components *d-element-A* (*deA*) and *d-element-B* (*deB*), or, the two types of pairs, i.e., "*d-elements*", which are shown in Fig. 10.2 and 10.3 – and which can be approximately described as a double helix, like in Fig. 10.4.

The terminal of a double helix – i.e., the double helix resulting from the superposition of two volleys of opposite polarity - can be either a "*d element*" of type *deA* or type *deB* – i.e., respectively, composed by a pair of superimposed "elements" κ_H and oriented either to one side or the other of the double helix propagation. In both cases, the terminal of one of the two helices refers to time or antitime.

For simplicity, we refer to the final terminal of every volley, even if the same goes for the leading terminal of every volley. In any case, the head of a volley in the α world is the tail of the same volley in the γ world, and vice versa.

We refer to the conventional representation of Fig. 10.3, instead of that of Fig. 10.2. In fact, once again we stress the completely arbitrary nature of every chosen graphic representation, as every representation refers to the internal structure of a photon, which is expelled by a "slingshot effect" from the atom monad, while the monad is observed from the outside. The empirical constraint prevents us from observing the *real* processes that take place within the atom monad. We can only base our knowledge on how much is expelled from the monad by the "slingshot effect".

That is, it should be emphasized that we must not confuse true physical objects - unknown *per se* due to the empirical constraint - with mere symbolic representations. Furthermore, we can assume (as above) that the full or empty blue (or light grey) arrows in Fig. 10.2 are characteristics of positive or negative particles (or the full-colored arrows pointing up or down). Equivalently, the same argument applies to antiparticles with reference to the symbols in red (or dark grey), instead of blue (or light grey).

Similarly, we can provisionally assume that torsion symbols are associated with the "spin" and/or "magnetic moment" of particles or antiparticles, but this will be clarified further below.

Summing up, the structure of a photon is therefore represented by Fig. 12.1, which is identical to Fig. 10.3. Note that, consistent with Fig. 10.4, we arbitrarily associated a counterclockwise twist with a blue (or light grey) arrow and a clockwise twist with a red (or dark grey) arrow. However, this agreement can be changed. In any case, it is important that the two twisting directions are opposite for particles and antiparticles.

Let us also suppose that we can symbolically associate a single torsion symbol associated with a single decoupled "*element*" with the "*spin*", and a pair of torsion symbols not associated with "elements" κ_E with the "*magnetic moment*".

To be more explicit, one should refer to the concrete examples of the figures shown in Chapter 13, according to Table 12.1, the meaning of which will be clearer after reading Chapter 13.

"*Spin*" and "*magnetic moment*" are intrinsic properties – which are now a "classic" in physics – and are characteristic of every particle or antiparticle.

These properties were introduced, historically, by physicists studying the interactions between particles.

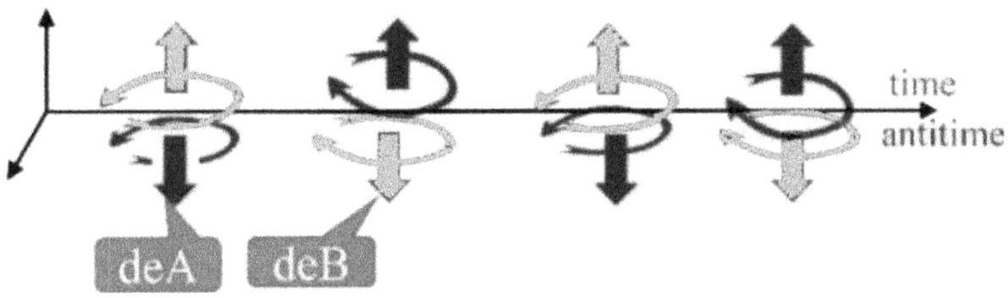

Fig. 12.1. Alternative representation of the same concept expressed in Fig. 10.2, through a different conventional graphic "language". That is, instead of empty arrows, we choose two symbols with a torsion symbol that better denote the mirroring of the κ_H *"elements"*. Particles are conventionally associated with blue (or light grey) symbols and antiparticles with red (or dark grey) symbols (see below). The association of the blue (or light grey) arrows with anticlockwise rotation is arbitrary. However, it is important that the direction of rotation is opposite between world α and world γ. The symbols *deA* and *deB* denote two types of coupling between *"elements"*, as will be better seen below. The acronym *"de"* stands for *"d-element"*, i.e., double *"elements"*, which can be of two types, *A* and *B*. This example shows the case of a photon that, according to the logical scheme adopted here, has no spin and no magnetic moment. See text.

Table 12.1 – Examples of various volleys with or without spin, with or without magnetic moment, oriented parallel or antiparallel to the spin

figure	*electric charge*	*spin*	*magnetic moment*
13.1	neutral (photon)	No	No
13.2	positive or negative	No	No
13.3	neutral particle or antparticle	No	No
13.4	positive or negative	No	Yes
13.5	positive or negative	Yes	No
13.6	positive or negative	Yes (with orientation opposite to the magnetic moment)	Yes
13.7	neutral particle or antparticle	Yes	No
13.8	neutral particle or antparticle	No	Yes
13.9	neutral *"Messenger"* of an electron or of a positron	Yes	No

To clarify the concept, let's remember the few notes made in Chapter 4 regarding the collision between billiard balls. We specified that, with Newton's principles of dynamics, the necessity is shown of the *conservation of energy and momentum*. Physicists have therefore assumed, very reasonably, that they can hypothesize that the same conservation laws also apply in interactions between particles.

Let us also remember that the most skilled billiard players know how to give a ball a spin on itself for the purposes of the game, which is transmitted to the struck ball causing special effects. This rotation effect is expressed - according to Newton's dynamics - as a further conservation law, called "*conservation of angular momentum*". Angular momentum takes into account the mass of the ball and its rotation speed.

Physicists have found that when particles or antiparticles interact they behave like billiard balls that have an intrinsic spin, a bit like rotating tops spinning in one direction or the other, and this property has been called "*spin*". But don't think that this conclusion was the result of "immediate" and "easy" thinking. It derived from hard thoughts and long discussions about "curious" experiments that appeared mysterious. It was a long and wonderful logical labor, which finally led to a shared conclusion.

A similar concern dealt with the *magnetic properties* of particles and antiparticles involved in an interaction. It appeared obvious to require that the magnetic properties summed among all the interacting objects (particles and antiparticles) must be conserved and identical to the magnetic properties summed among all the objects (particles and antiparticles) produced after the interaction. However, to be confirmed, this further presumed conservation law implies that every particle or antiparticle must be associated with a well-defined intrinsic "*magnetic moment*". In other words, particles and antiparticles look like rotating balls or tops, which, in addition, are like rotating magnets. This new property (intuitively like a rotating magnet) is called the "*magnetic moment*" of a given particle or antiparticle, respectively. All this is the result of mere observations, suitably interpreted in the most reasonable way possible, carried out and discussed over the course of decades, and which has long been unanimously shared by all scholars.

Mother Nature is indeed curious! Particles and antiparticles behave like billiard balls that are magnetized and have an intrinsic spin. It is mere and unequivocal evidence observed in Nature. There is nothing "metaphysical" or speculative. This is something very clearly shown by experiments, and now unanimously acknowledged by all specialists and for many decades. It is an incontrovertible outcome. However, due to the empirical constraint, this conclusion was achieved only after a lively debate among the most enlightened minds, discussing numerous, very delicate - and very expensive - experiments carried out with elementary particles.

Particles and antiparticles, therefore, have intrinsic properties that are generically referred to as "*quantum numbers*". But the story does not end there. Other "*quantum numbers*" have been further hypothesized to explain new "*conservation laws*" aimed to empirically justify further experimental evidence. The result was a "zoology" of elementary particles that was both very complicated and fascinating.

The ensemble of new "*quantum numbers*" seems almost never-ending ("*strangeness*", "*flavor*", "*charm*", etc.). They are empirical concepts, which are observed and appear necessary to explain why certain interactions occur between particles and antiparticles, unlike others. In reality, no one knows their meaning, other than that it is something that is necessary to explain some regularity shown by nature. A new discipline was thus born, called "*quantum chromo-dynamics*" (*QCD*).

We note that "*quantum numbers*" are conceived as intrinsic properties of particles and antiparticles - just like the intrinsic "*virtues*" hypothesized by Aristotle in antagonism to the principle of inertia, as recalled in Chapter 4.

For comparison, we also recall here the development - that has been, and is still being, much discussed and studied – connected to the hope of being able to revive the splendor that the Schrödinger equation had in relation to the atom monad. That is, an attempt is made to generalize the same formalism of the Schrödinger equation, extending its application to the dynamics of particles and antiparticles. This gave rise to a very complicated discipline, the so-called "*quantum electrodynamics*" (*QED*), which is considered among the most advanced frontiers of theoretical physics today. The results are still very partial and subject to heated debate.

Among other things, we recall a *QED* paradox that has been well known for many decades. In fact, *QED* is treated with *continuous* functions, sometimes expressed as an infinite series of terms. A series of terms - which has physical meaning - must have, when concretely added, a sum that does not diverge at infinite. This is the same paradox as the "spectrum" of the "black body" mentioned in Chapter 2. Max Planck managed to resolve that disturbing paradox by hypothesizing a phenomenon described with discrete quantities ("quanta") as opposed to a treatment strictly carried out with continuous functions. His idea was truly a "lightning", and at that time, it was rowing against the current, it was disruptive!

Similarly, in *QED*, if the individual terms of an infinite series of terms are considered, every term is interpreted as an elementary process that resembles a more or less complicated game of collisions between billiard balls of various kinds. These are the so-called "*Feynman graphs*", which were a "lightening" idea of Richard Phillips Feynman (1918-1988). However, by adding the terms of the entire series, the final result is absurd, as the sum is infinite. So scientists talk about the need to review the theory to avoid this paradox - and they therefore talk about the need for a "*renormalization*" theory, which is a topic on which rivers of ink have been spent by the most brilliant minds.

In fact, just consider that - as in the case of Planck - if we avoid using the illusion of "continuity", the number of terms to be added is certainly no longer infinite. Instead, everything is discreet. Therefore, the physically significant terms in the approximation with continuous functions must be bounded, and gives a finite result! In other words, the paradox no longer exists! That is, the logical "bug" of the illusion of "continuity" is the origin of this paradox.

QED and *QCD* do not require further details here, also because they still require a lot of thought and debate - and a comparison with the enormous set of very advanced experiments already available and which are continually expanding. *QED* and *QCD* constitute a true "monument" of cutting-edge research, and it is difficult - if not impossible - to make a brief summary of it, even by a specialist involved in the work.

In our logical journey, we stress that only the approach proposed here allows a reasonably "simple" framework, free of paradoxes. But the road ahead is still very long.

Please also remember, as is well known, that these are not merely academic disquisitions. In fact, understanding these phenomena can open the way to the legendary exploitation of nuclear fusion energy, i.e., that form of energy found on the stars and which man has only been able to reproduce in the form of a hydrogen bomb. For many decades, almost for a century, the dream of theoretical physics has been to provide a logical interpretation that allows the process of nuclear fusion to be reproduced in a controlled manner, and thus to find a source of clean energy, without radioactive waste, and in quantity unimaginable until now. However, to achieve a sufficient understanding of the phenomena, it is essential to first get to grips with the paradoxes.

We can now move on to consider in a little detail the "zoo" of particles and antiparticles - and how the spiral structure derived from that of the photon considered as a double volley of miniphotons can explain the observed phenomena. Our path becomes increasingly harder, more intricate, but increasingly fascinating…

Chapter 13

Particles and Antiparticles – "*Lame*" Volleys

"the test of a person is in conversation ..."

[*Sirach* 27, 5]

....

In Chapter 12 we clarified how particles and antiparticles are part of a "zoo" of experimental evidence - still constantly expanding, due to increasingly complicated experiments performed with ever more powerful particle accelerators. The discipline today known as *QCD* was born from this, while the path followed by the most advanced theoretical physics is trying to revive the success of the Schrödinger equation for the atom, and thus *QED* was born.

We stress, however, the essentially arbitrary - undoubtedly reasonable - contents of this entire formulation, based on the assumption of *continuity*, and on the formalism inspired by the Newtonian model with forces, energies, masses, instantaneous action at a distance, etc. The purpose of our journey is now to identify, if possible, the causes that imply the paradoxes that afflict today's understanding of the subatomic world.

To this end, we will have to introduce imaginative axioms, i.e., working hypotheses – and in any case, even the formulation which is based on an imitation of the Newtonian conception implies *a priori* assumptions which, ultimately, are equally arbitrary, although supported by the merits acquired by engineering applications in the macro-world in which we live.

But let's go in order. We faithfully reproduce some considerations already anticipated in Chapter 12 when we introduced the concept of "*spin*" and "*magnetic moment*" of a particle or antiparticle.

The quantum and spiral structure of a photon should be conceived as a volley of miniphotons. When the discrete structure is interpolated with a continuous function, we find a double spiral, where one spiral is in time and one is in antitime.

Let us consider the quantum and spiral structure of particles and antiparticles, as it follows when we move from the β world to the α and γ worlds.

Let us consider the components *d-element-A* (*deA*) and *d-element-B* (*deB*), i.e., the two types of pairs, i.e., "*d-elements*", shown in Fig. 10.2 and 10.3, which can be roughly described as a double helix as in Fig. 10.4. The terminal of a double helix – i.e., superposition of volleys of opposite polarity - can end with a "*d element*" of type *deA* or type *deB*, respectively, or with the pair of "*elements*" κ_H superimposed and oriented either to one side or to the other of the propagation of the double helix. In both cases, the terminal of one of the two helices refers to time or antitime, respectively.

For simplicity, we refer to the final terminal of each volley - even if the same goes for the leading terminal of each volley. The head of a double volley in world α is the tail of the same double volley in world γ, and vice versa.

We refer to the conventional representation of Fig. 10.3, instead of that of Fig. 10.2. We once again repeat the completely arbitrary nature of every chosen conventional graphic representation, since we refer to the internal structure of a photon - which is expelled by "slingshot effect" from the atom monad while the monad is observed from the outside. Therefore, the empirical constraint prevents us from observing the real processes that occur within the monad. We can only "construct" our knowledge based on what is expelled from the monad through the "slingshot effect".

That is, it should be stressed that we must not confuse real physical objects - in themselves unknown due to the empirical constraint - with mere symbolic representations that we arbitrarily define for mnemonic "simplicity". Furthermore, we can assume (as above) that the solid blue (or light grey) arrows in Fig. 10.2 are characteristics of positive or negative particles (for the arrows oriented up and down, respectively). Likewise, we can tentatively assume that the torsion symbols are associated with the "spin" and/or "magnetic moment" of particles. As far as antiparticles are concerned, refer to the symbols represented in red (or dark grey), with similar conclusions, although with the necessary distinctions.

In summary, the structure of a photon is depicted by Fig. 13.1, which is identical to Fig. 10.3 and Fig. 12.1. Note that, consistent with Fig. 10.4, we arbitrarily associated a counterclockwise twist with a blue (or light grey) arrow and a clockwise twist with a red (or dark grey) arrow. However, this convention can be swapped. In any case, it is essential that the two directions of twisting are opposite for particles and antiparticles.

Let us also suppose that we can symbolically associate a single torsion symbol of an "*element*" κ_H with the "*spin*". Furthermore, let us suppose that we can symbolically associate a coupled pair of torsion symbols of "*elements*" κ_H with the "*magnetic moment*". Concrete examples are shown in the figures below.

Please remember again, however, that this is only a *provisional symbolic and conventional graphic representation* of some concrete physical entity that has an inherent nature unknown to us. Nor the symbolic interpretation proposed here necessarily is the "best" possible. Alternative proposals may possibly be made and are welcome. Consider that other "*quantum numbers*" besides "*spin*" and "*magnetic moment*" – such as "*strangeness*", "*flavor*", "*charm*", etc. (see Chapter 12) - remain to be represented symbolically if necessary. That is, it all requires a long study - with reflection and discussion - which has not yet been carried out. Ultimately, this is the job of *QCD*.

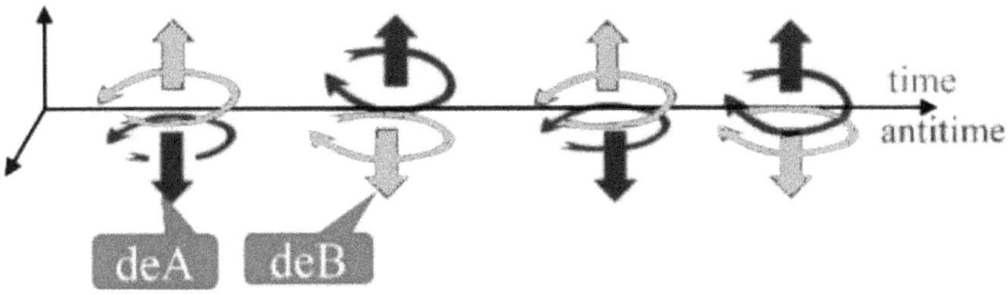

Fig. 13.1. Alternative representation of the same concept expressed in Fig. 10.2 and 12.1, through a different conventional graphic "language". That is, instead of empty arrows, we choose two symbols with a torsion symbol that better denote the mirroring of the κ_H "*elements*". Particles are conventionally associated with blue (or light grey) symbols and antiparticles with red (or dark grey) symbols (see below). The association of the blue arrows with anticlockwise rotation is arbitrary. However, it is important that the direction of rotation is opposite between world α and world γ. The symbols *deA* and *deB* denote two types of coupling between "*elements*", as will be better seen below. The acronym "*de*" stands for "*d-element*", i.e., double "*elements*", which can be of two types, *A* and *B*. This example shows the case of a photon that, according to the logical scheme adopted here, has no spin and no magnetic moment. See text.

Let us now introduce the concept of a "*lame*" volley due to the fact that - in comparison with an intact volley as contained in the double spiral of the photon - the volley has lost some piece at one or the other of its terminals.

Let us therefore consider the possible cases of "*lame*" volleyballs (Fig. 13.2, ..., 13.9). We must then foresee the possible association of real observed entities with every different type of symbolic representation of a "*lame*" volley. On the other hand, we underline that *a priori*, no physical reason requires that *every possible* case study of "*lame*" volley must have a physical correspondence with what is currently known and observed in reality. A synthesis has been presented in Table 12.1.

The case of particles or antiparticles with electric charge and no spin and no magnetic or antimagnetic moment is shown in Fig. 13.2.

What is the meaning of superimposed positive/negative particle/antiparticle with no spin and no magnetic moments (Fig. 13.3)?

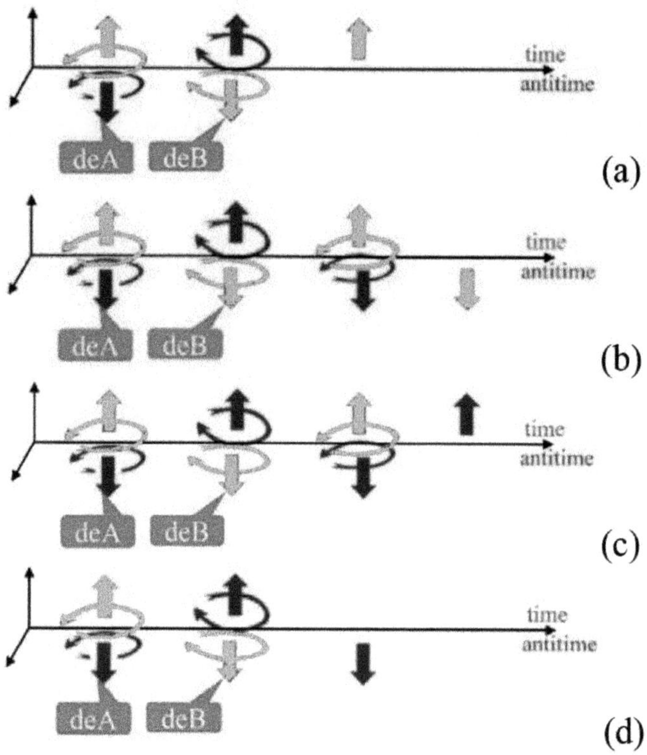

Fig. 13.2. Positive (or negative) particle (or antiparticle) with no spin and no magnetic moment. See text.

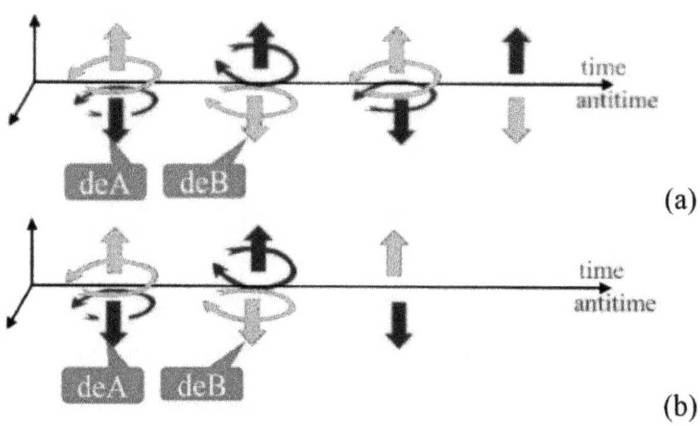

Fig. 13.3. Neutral particle/antiparticle with no spin and no magnetic moments.

What is the difference between a positive (or negative) particle (or antiparticle) with magnetic moment and no spin (Fig. 13.4) and the same one with spin and no magnetic moment (Fig. 13.5)?

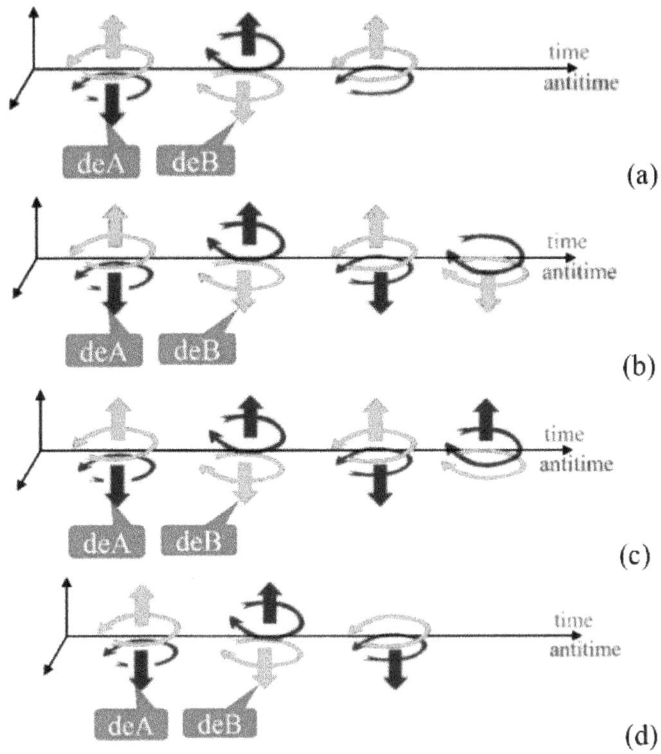

Fig. 13.4. Positive (or negative) particle (or antiparticle) with magnetic moment and no spin. See text.

A reasonable assumption is that an electron should be represented by a case associated with Fig. 13.5 or 13.6. We stress that in either case, no asymmetry characterizes the electron (and the positron).

In fact, some authoritative scholars have emphasized that if the Universe had been perfectly symmetrical, there would be nothing left but light, i.e., the β world alone. That is, these scholars fear that in a perfectly symmetrical world – i.e., with an equal quantity of matter and antimatter - a perfect symmetry between the α world and the γ world would lead to a complete annihilation in photons, therefore a total disappearance of the α world and the γ world which they would flow into the β world. In fact, this is what happens in the already mentioned cycle of galaxies of matter, and of antigalaxies of antimatter, generated by the materialization of photons (see Chapter 4 and elsewhere *passim*). These galaxies, once their cycle is completed, annihilate themselves, to then regenerate - through materialization - new galaxies and antigalaxies, etc. Therefore, there is always symmetry between the observations of matter and antimatter in the Universe. In this respect, see also the discussion on superluminal phenomena reported in Chapter 22, in Chapter 23, in Chapter 24 and in Chapter 25. That is, we should not be surprised if we observe more matter than antimatter, given that this only depends on the empirical constraint and on the fact that we live within a galaxy in world α.

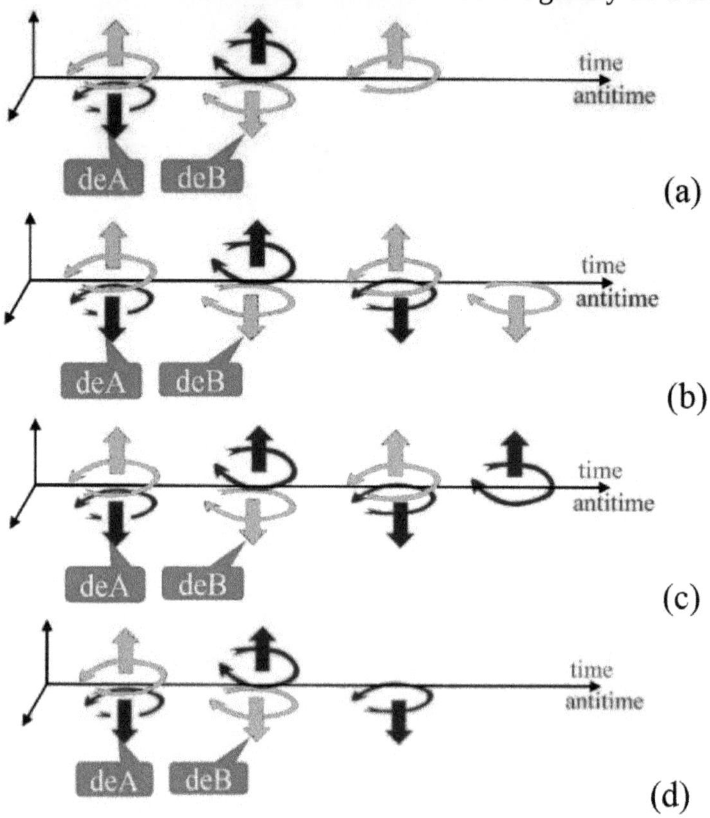

Fig. 13.5. A positive or negative particle or antiparticle with spin and no magnetic moment. See text.

What is the meaning of a positive (or negative) particle with a magnetic moment of opposite polarity (Fig. 13.6)?

Furthermore, we also remember the perpetual search for equilibrium in the Universe, according to the golden ratio (Chapter 7), and we remember the irrationality of the golden ratio φ, which guarantees the *irreversibility* of the arrow of time and antitime.

This property is essential precisely for the ability to perceive "time" and the direction of the arrow of "time" - both by man and by all living beings. Furthermore, as illustrated in Chapter 27 and in Chapter 28, this perception is fundamental for survival also in terms of autopoiesis, i.e., the generation of new individuals, aimed at replacing the elderly and dying individuals, according to "statistics" and antifragility.

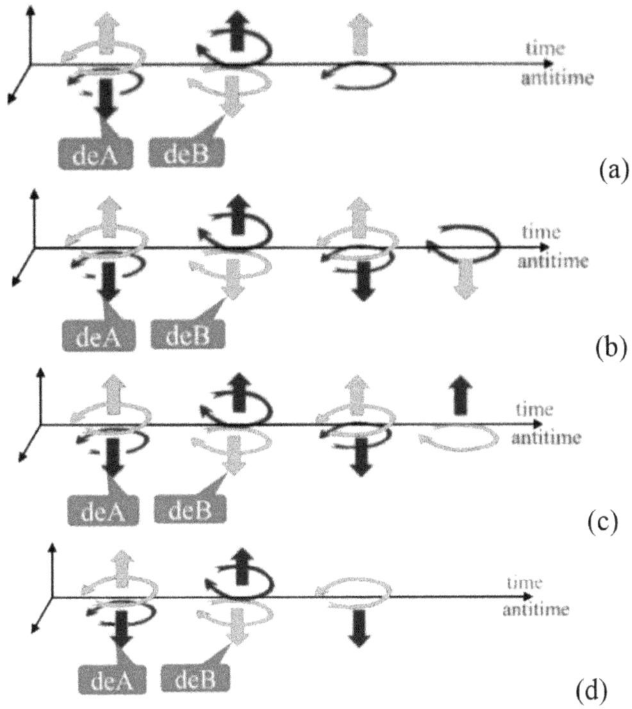

Fig. 13.6. A positive or negative particle or antiparticle with no magnetic moment and with spin of opposite polarity. See text.

Returning to our schematic, what does it mean when the blue (or light grey) and red (or dark grey) arrows survive together with only one spin and no magnetic moment (Fig. 13.7)?

Fig. 13.8 represents a neutral particle or antiparticle with a magnetic moment. Can a neutral particle have a magnetic moment if it has no magnetics? However, a neutral particle or antiparticle without a magnetic moment is simply a photon, like Fig. 13.1, although possibly with fewer pairs *deA* and *deB*.

On the other hand, what does a "*lame*" photon mean due to the lack of just one *deA* or just one *deB*? That is, when both κ_E "*elements*" are missing, while both κ_H "*elements*" survive (Fig. 13.8), one can imagine that this is (perhaps) the case of a neutral particle with a magnetic moment. But one wonders if this is perhaps a special case without physical reality, which is just a possibility derived from complicated mathematical calculations.

Please also note that, in listing the possible cases, the case of a neutral particle or antiparticle with spin and magnetic moment was not foreseen (see Table 12.1). It would be like the case of Fig. 13.7 and Fig. 13.8, but "*doubly lame*", which seems impractical and unrealistic. That is, if this is true, there should exist no particle or antiparticle of neutral electric charge and having both spin and magnetic moment.

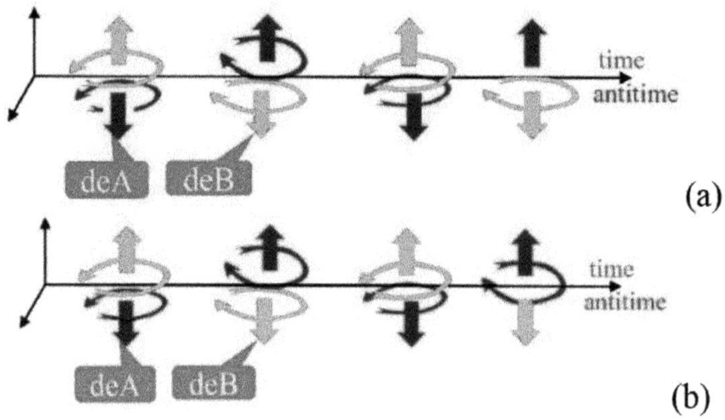

Fig. 13.7. Neutral particle/antiparticle combined with only one spin and no magnetic moment. See text.

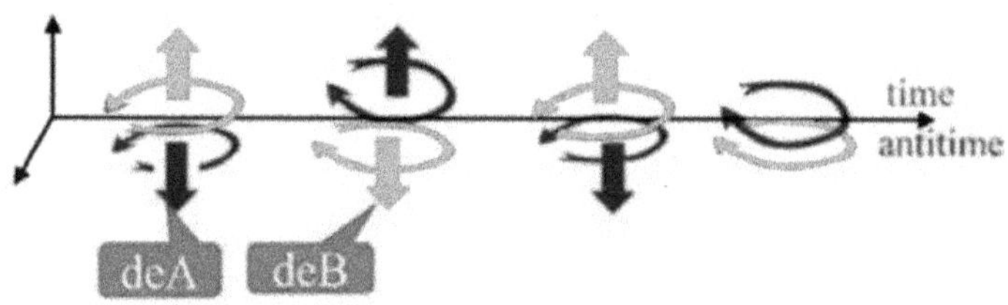

Fig. 13.8. Neutral particle (or antiparticle) with magnetic moment and no spin. See text.

Obviously, in this game of ours with a sort of "abacus", anything is possible. Just hypothesize about new assumptions, etc. All of this remains to be clarified with further comparisons with the "zoo" of particles and antiparticles that are observed.

In any case, note that, in this way, the distinction between a neutral particle and neutral antiparticle is only a matter of convention, since the distinction depends only on whether the object is observed in the α world, or in the γ world, while in reality, the object is the same even when observed in different worlds. These cases – i.e., the meaning of neutral particle and neutral antiparticle, including the connection to gravitation – are discussed in Chapter 15.

On the other hand, Fig. 13.9 can represent, perhaps, a neutral particle or antiparticle with spin. In reality, these are probably typical "*messengers*" which - according to the *SEA* discussed in Chapter 14 - are associated respectively with an electron or a positron (see below). Note that a "*messenger*" should not be confused with the electron (or any other source) that emits the "*messenger*". That is, the electron (or positron) has a charge, whereas a "*messenger*" has no charge.

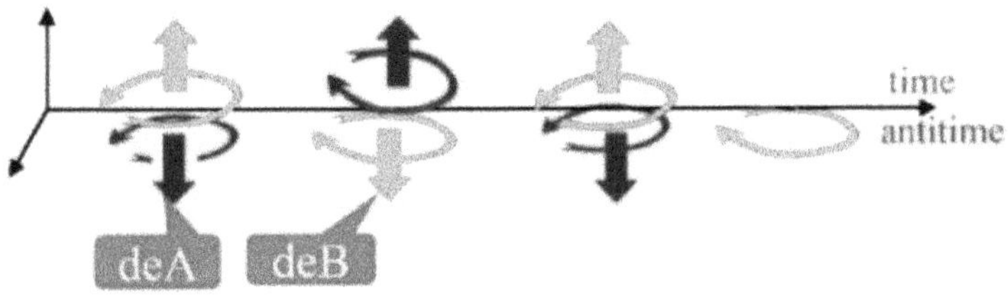

(a)

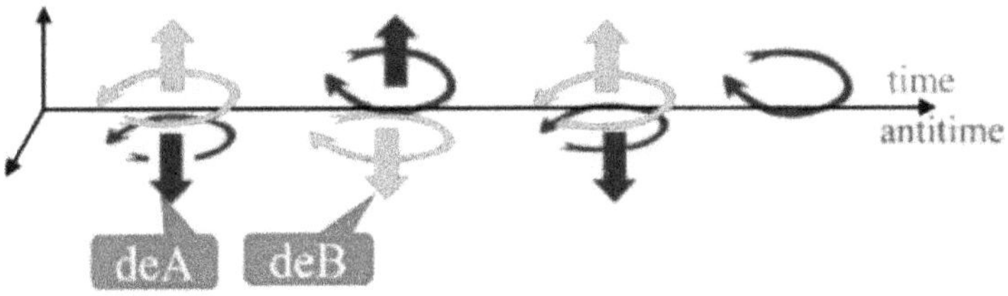

(b)

Fig. 13.9. "*Messenger*" of an electron or positron with neutral charge, with spin, and without magnetic moment. See text.

A possible hypothesis is that – according to an analogous criterion – a neutron could be a spinless "*messenger*", i.e., the question is whether Fig. 13.1, or 13.7 or 13.8 can represent a "*messenger*". Fig. 13.1 is what we observe and call a "photon," with no spin and no magnetic moment. In principle, the propagation of a photon requires no spin and no magnetic moment. In fact, a photon needs to be suitable only for the dynamical equations derived in Chapter 20 and in Chapter 21.

The distinction between Fig. 13.9 and Fig. 13.7 or Fig. 13.8 is more uncertain. Note that the surviving components can both be κ_H "*elements*" and can be part of either one *deA* or one *deB*. We can't know whether or not this makes a physical difference.

The "*element*" κ_E symbolically represents the positive or negative charge of the particle that justifies the electric interaction of the particle, according to the equations of motion formally defined in Chapter 20 and in Chapter 21. The "*element*" κ_H - or torsion symbol - can be associated with the spin and the magnetic moment of the particle or antiparticle, and - as shown in Fig. 5.3 - can be associated with what, in classical physics, is said to be the magnetic field of an electric current. This concept will be further discussed in Chapter 14.

That is, we emphasize the competitive role between the "electrical" coupling due to the κ_E "*elements*" and the magnetic coupling due to the κ_H "*elements*".

Concerning the stability of a photon, the most important contribution is probably given by the pair of magnetic κ_H *"elements"* (blue or light grey and red or dark grey) that overlap. Their superposition therefore acts as two elementary magnetic poles, which strengthen the stability of the double helix. That is, even when both κ_E *"elements"* within a *deA* or a *deB* are missing, the surviving pairs of κ_H *"elements"* constitute an efficient *"glue"* that contributes to the stability of the double helix (see Chapter 10). When stability is lost, the volley decouples and materializes into particles and antiparticles of opposite charge and polarity, i.e., matter and antimatter.

Observations of antiparticles show that antimatter unquestionably exists - and indeed the existence of antimatter is one of the paradoxes encountered by the standard formulation of physics (Chapter 4). A summary of the possible cases is shown in Table 13.1 (but also see Table 12.1).

Table 13.1 – *"Lame"* Volleys and possible suggested physical association

Missing component in volley	*Resulting object type*
deA and *deB* or *deA* or *deB*	Lower energy photon Neutral object (i.e., photon or neutral particle or antiparticle?)
only one *"element"* κ_E	Positive (or negative) particle (or antiparticle) with magnetic moment (Fig. 13.4)
an *"element"* κ_E plus the *"element"* κ_H of opposite parity	Positive or negative particle or antiparticle with no magnetic moment and with spin of opposite polarity (Fig. 13.6)
an *"element"* κ_E plus the *"element"* κ_H of the same parity	Positive or negative particle or antiparticle with spin and no magnetic moment (Fig. 13.5)
one "element" κ_E plus two *"elements"* κ_H	Positive (or negative) particle (or antiparticle) without spin and without magnetic or antimagnetic momentum (Fig. 13.2)
one *"element"* κ_H	Neutral particle or antiparticle with a single spin and no magnetic moment (Fig. 13.7)
two *"elements"* κ_E	Neutral particle (or antiparticle) with magnetic moment and no spin (Fig. 13.8)
two *"elements"* κ_E plus one *"element"* κ_H	*"Messenger"* of an electron or a positron with spin (Fig. 13.9)
two *"elements"* κ_H	Neutral particle/antiparticle with no spin and no magnetic moments (Fig. 13.3)

Therefore, a most energetic photon is a volley with a larger number of pairs of κ_E and κ_H "*elements*".

When annihilation between particle and antiparticle occurs, the result is a photon, which cannot be confused with a neutron or a neutral particle. In fact, a photon is "*undegraded*" emp, while a neutron or an antineutron is composed of "*degraded*" emp and cannot be confused with "*undegraded*" emp.

Therefore, as far as a photon is concerned, a volley with a greater number of pairs of "*elements*" κ_E and κ_H has a greater probability - due to greater instability - that decoupling occurs, i.e., a possible materialization in a couple particle/antiparticle. The kinetic energy of the particle and the antiparticle occurs according to the laws of motion formally implemented in Chapter 20 and in Chapter 21.

Therefore, a most energetic photon is a volley with a larger number of pairs of κ_E and κ_H "*elements*".

Conversely, a less energetic photon probabilistically has less instability and does not materialize. However, for the moment, it is not clear whether we should assume that, by losing an "*element*" κ_E, the volley must also lose the companion "*element*" κ_H, or not. This uncertainty will be considered below, because a decoupled "*element*" κ_H seems essential to justify Faraday's law of electromagnetic induction (see Chapter 14) and beyond (see Chapter 15). This appears to be a crucial aspect, but - unless there is a pressing need - we must refrain from hypothesizing further speculation.

Consider, however, that all the cases mentioned above refer to "*lame*" volleys which - in space-time - lack some "*element*" in the "*head*" terminal or - symmetrically in antitemporal space - in the "*tail*" terminal of the volley. However, all of the above arguments hold regardless of whether the missing "*elements*" are at the "head" or "tail". This is an additional degree of freedom for physical interpretation.

In any case, any specific interpretation proposed here is necessarily speculative – and we must refrain from adding any further arbitrary speculation, unless prompted by observational evidence that cannot be explained more simply and with no paradox.

We have thus completed a brief discussion of the possible case studies of "*lame*" volleys and their respective associations with the phenomena actually observed. It was like playing with an abacus. Let's now discuss how particles and antiparticles, structured according to the pattern of "*lame*" spirals, can interact with each other, generating new particles and/or antiparticles. Our "game" becomes increasingly complicated…

Chapter 14

Particle, Antiparticle, and Photon Interaction – *SEA*

"No disciple is superior to the teacher; but when fully trained, every disciple will be like his teacher."

[Gospel of Luke 6: 40]

"Tristo è lo discepolo che non avanza lo suo maestro"

[*"Sad is the disciple who does not surpass his master"*]

Leonardo da Vinci (1452-1519) (Cod. 2293)

....

We now consider how particles, antiparticles and photons - with their respective spiral structures - interact with each other. To this end we introduce an important additional hypothesis, or axiom.

SEA (Steady Exhalation and Annihilation)

Every sample of "emp" – i.e., everything that "exists" – constantly releases photons of both parities. However, in world α only photons of standard parity can be observed, while in world γ only antiphotons of the opposite parity can be observed.

In both cases, both a photon and an antiphoton can be detected *only in the case* where the photon is captured by a *receiver*. If no *receiver* is available, the photon or antiphoton "dies" in absolute space, almost like something that never existed. In fact, it is never captured and "observed" by anything.

The constant release of photons/antiphotons from every "emp" sample is called *SEA* (*Steady Exhalation and Annihilation*). These photons/antiphotons should be considered as "*messengers*" that we cannot see or detect, at least until they produce any effect, as explained below. Moreover, to explain the paradox of the Poynting vector, photons of both parities must be considered (Fig. 14.1, already shown as Fig. 6.2).

The associated detailed analysis leads to the definition of "monad" (see Chapter 6). In other words, every natural system is based on phenomena that occur within a "domain" of spacetime called "monad" which is limited and confined. This is a general rule. It is noted - as an observed matter of fact – that, in Nature, reality seems to seek compact structures - hence the property of "compactness" aimed at occupying the widest possible "domain" in spacetime, through a "repetitiveness" of the same structures in different formats and spatial and temporal dimensions.

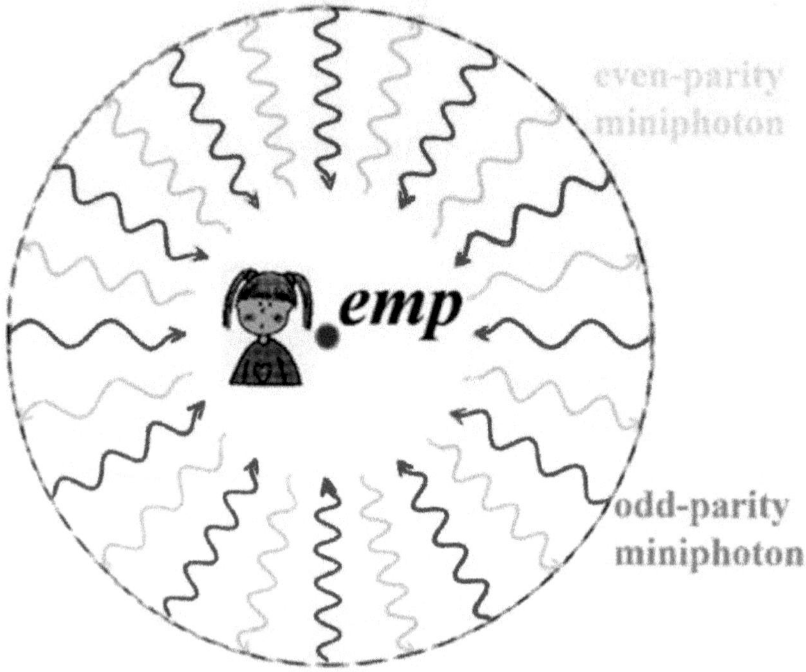

Fig. 14.1. Alice is the observer in the middle of the figure. Alice's "*luminous sphere*" denotes the spatial extension that an electromagnetic wave reaches during a time span Δt. This is a sphere of radius $R = c\Delta t$, where c is the speed of light. The figure shows the definition of "*monad*" in which Alice is immersed (see text). The Poynting vector that propagates in 3D space at the speed of light c is indicated by a blue (or light grey) wave propagating outward (see Chapter 5). The returning Poynting vector is shown by a red (or darker grey) wave propagating inward. The blue (or light grey) wave is composed of "miniphotons" of even parity, and the red (or darker grey) wave of "miniphotons" of odd parity, i.e., "antiminiphotons", i.e., with the arrow of time inverted. Every emp element always releases - as an endemic need - a sort of "*rubber band*" (or a "*double helix*") composed of photons of both parities. A formal explanation is in terms of the *SEA* axiom and the double helix. See text.

Fractality and the Golden Rule are the fundamental properties that characterize the evolution of all natural phenomena (see Chapter 7). The irrationality of the golden ratio φ is fundamental, since it implies that phenomena never reach final equilibrium. The phenomena are rather in continuous search for a final equilibrium which, however, can never be reached, since the ultimate state of equilibrium is continuously redefined and updated with the evolution of the Universe. In the final analysis, this is the true meaning of the perception of the flow of time.

In fact, this crucial property is identified in the world α (see Chapter 9), with the perception of "time" and the direction of the arrow of "time" - while in the world γ a symmetric property holds that concerns antitime.

In other words, the irrationality of φ is the *necessary* condition for our ability to perceive "time" and "antitime" and their respective arrows – i.e., the irrationality of φ is intrinsic to the perception of "*irreversibility*". This holds for man, and for every living being, since this is an essential prerequisite for survival, even if - perhaps - only man has memory and awareness of "time".

This concept is crucial (see Chapter 28) even when we talk about the survival of life, which is opposed - endemically - to the natural aging of every being, through self-reproduction, i.e., the generation of new young individuals who replace the old ones (autopoiesis).

A related additional constraint concerns the impossibility - for a strict matter of principle - to refer to subsystems of the natural reality of the entire Universe. Indeed, the human mind wants to refer to "simple" laws, applied to natural subsystems. That is, human beings - in accordance with Galileo's teaching - seek chains of cause and effect, i.e., they seek "*linear*" systems.

This approximation is certainly very useful for application purposes, even if it cannot include "*irreversibility*". Indeed, all the "laws" thus deduced by humans seem invariant with respect to time reversal - and this is considered a true paradox, even in most advanced physics. However, time's *reversibility* is not physical, Nature is *irreversible*! The temporal reversibility of laws conceived by man derives from the assumption that a law expresses a relationship between *equilibrium states* of the natural system, while in reality, *nothing is in equilibrium*.

In other words, if we want to deal with a rigorous and abstract formulation, we must refer to the entire Universe as a single system, which also includes worlds of antimatter. On the contrary, we can consider subsystems only as long as we allow approximations and recognize our inability to include the arrow of time and irreversibility in partial laws. "*You cannot pick a flower without disturbing a star*" (Galileo).

Therefore, the perception of "time" is closely linked to the ability to observe the whole Universe. The surprises of our logical journey never end …

The dynamics of the interaction among particles, antiparticles and photons

The formal definition of "mass" (or more properly of the emp content) of a particle (or antiparticle) depends on the total length of the volley of "*d-elements*" (see Chapter 13), as every addition of new "*d-elements*" acts as a "*booster*" (i.e., as a sort of additional rocket) which changes the dynamic behavior of the object.

The classical dynamic parameters - such as mass and kinetic energy - are defined through the formal calculation of the dynamic equations according to Newton's principles, and according to the laws of conservation of energy, momentum, and angular momentum (see Chapter 4).

The conservation law of emp is the equivalent of the law of conservation of energy, and is correctly expressed in terms of "quanta" – because the "energy" of a photon, or the "mass" of a particle or of an antiparticle, are expressed by the number of "*elements*" that compose the respective volley.

On the contrary, when we talk about conservation of momentum, we are referring to the product of the mass - which is expressed in "quanta" - and the velocity, which is a vector defined in the α world or in the γ world, respectively, by "time" and "antitime". That is, this vector is a continuous quantity, therefore it represents something continuous like "time" or "antitime".

Furthermore, due to a need for self-consistency - expressed by the "*ABC theorem*" (see Chapter 3) - it can be shown (see Chapter 20 and Chapter 21) that in our rigorous formulation it is strictly necessary to refer, at most, to a quantity of *generalized* momentum, expressed as the product of mass and speed multiplied by an appropriate weight function. In any case, it must be stressed that an approach of this type has paradoxical logical contents, since it implies reference to a continuous, i.e., non-quantized, entity.

On the other hand, *generalized* momentum is a "statistical" concept, as it is derived from dynamic equations, and is the result of a "continuous" approximation, i.e., it is a "statistical" description of an intrinsically quantized phenomenon.

In contrast, when we deal, e.g., with the interaction between a particle and a photon – as in Compton scattering (see Chapter 15, Chapter 18 and Chapter 21), or as in the case they are billiard balls in 3D space – we must consider that after the interaction some non-"*lame*" segment of the incoming photon detaches from the photon and remains attached to the particle or antiparticle, just like a "*booster*" - i.e., a sort of additional "rocket" that changes the speed of particle or antiparticle. This explains the changed "kinetic" energy of the particle.

However, the representation of the phenomenon in terms of continuous functions, such as momentum or generalized momentum or kinetic energy etc., is only a way to interpolate - by means of a continuous function - onto a quantized scheme which, conversely, is intrinsically discrete, like Feynman graphs (see Chapter 5 and Chapter 12).

In other words, the calculation of the dynamics of particles and antiparticles must be carried out by counting the "*elements*" of the miniphotons, and considering whether the involved volleys are "*lame*" or not (see Chapter 20 and Chapter 21).

It should be stressed that this distinction is *not* a subtle mental lucubration. Indeed, consider that Max Planck got rid of the paradox of the black body spectrum by abandoning interpolation using continuous functions, and by referring directly to quantized phenomena (see Chapter 2). Therefore, the quantized description expressed in terms of "*booster*" that is attached to the interacting particle is correct.

The conservation law of emp is consistent with the founding axioms of our journey. On the contrary, the formal continuous equations of dynamics, derived from the law of conservation of generalized momentum, are approximate "statistics", i.e., appropriate interpolations averaged to describe the otherwise quantized structure of phenomena.

Similar considerations also apply to the law of conservation of generalized angular momentum, but - as we will see - this is only a problem of technicalities which add nothing to the problems of principle illustrated above.

When the kinetic energy is zero (calculated with respect to a chosen reference frame), the mass is called the "*rest mass*" of a particle or antiparticle. However, this is true when the mass is referred to the given reference system that has been arbitrarily chosen. That is, in principle, one should only refer to the "absolute" reference system, which certainly exists. On the other hand, we cannot discuss the rest mass referring to the "absolute" reference system, i.e., the æther, if we do not know how to recognize, define, and observe the æther.

The "mass" of a particle or antiparticle represents a sort of "inertia" that slows down - or delays - the mobility of the object. Let's remember the classic discussion on "*inertial mass*", as opposed to "*gravitational mass*" (see Chapter 5). In fact, it is observed that a "*lame*" volley moves much slower than a complete volley, which instead moves at the speed of light c, both in the α and in the γ world.

That is, a "*lame*" volley is a "*degraded*" form of emp, compared to the "*undegraded*" form, which is represented by a complete volley, i.e., by a photon. A "*lame*" volleyball is like a runner without a leg.

Furthermore, it is observed that electrons are much more mobile than protons, therefore, it is stated that electrons have a much lower mass than that of a proton. Therefore, the electric currents considered in classical physics, of Newtonian and Maxwell derivation, are composed exclusively (or almost exclusively) of the motion of electrons - and to a much lesser extent also of the motion of protons and ions. Therefore, due to their greater mobility, electrons constitute a sort of "*less degraded*" form of emp, and play a leading or exclusive role in phenomena of electrical conduction.

Consider an electron represented by a "*lame*" volley as it has a terminal "*d-element*" which lacks the positive "*element*" κ_E while only one "*element*" κ_H survives as in Fig. 13.5b (or perhaps two "*elements*" κ_H as in Fig. 13.4b). Consider a surviving κ_H "*element*" or two surviving "*unbalanced*" κ_H "*elements*" associated with the "*lame*" electron volley. Consider another electron flowing within a circuit, and consider the *SEA* mechanism. We must consider two opposite actions.

The first action is manifested as an electrical "repulsion" between the two electrons (in agreement with the classical Coulomb law; see Chapter 5) – according to the logic of the dynamic equations illustrated in detail in Chapter 20 and Chapter 21. In this respect, however, consider that the dynamic behavior of an electron (and a proton) depends on the reciprocal completion of the two respective "*lame*" volleys. In fact, in the case of "*attraction*" the missing "*elements*" compensate each other, and thus a new complete volley is formed which can move at the speed of light. In the case of failure to compensate, there is "*repulsion*" between particles with an electrical charge of the same sign. That is, the case of the interaction between two electrons – one stationary and the other moving – can experience nothing other than "*repulsion*" if the "*messenger*" is of the type of Fig. 13.9a. We should rather consider the case of Fig. 13.5b. It is therefore necessary to develop more complicated models.

In any case, one can always imagine an alternative symbolic representation of quantized observational parameters – which have an unknown true physical nature in any case. Indeed, there is no "optimal" choice of symbolic representation. Every representation is "correct" to the extent that it meets the requirements of available observations. Ultimately, the "simplest" axiomatic definition is chosen according to Ockham's razor.

Ockham's razor, or *principle of economics*, establishes that among the various possible interpretative schemes or models, the simplest model must always be chosen, since it is probabilistically least influenced by human arbitrariness. William of Ockham (1288-1347), also known as William of Occam, is one of the most eminent figures of the important Franciscan theological and philosophical school, born in Oxford shortly after the death of St. Francis. William of Ockham was known as the *Venerable Inceptor* and *Doctor Invincibilis*. He had a troubled existence for having defended principles that are now universally shared, but which then seemed contrary to established authority. His vicissitudes - very freely reinterpreted - inspired the famous novel "*The Name of the Rose*" by Umberto Eco.

In our logical journey for the moment, we can hypothesize, e.g., that the *SEA* mechanism occurs - perhaps - through a mixture of different types of "*messengers*" - while the "*electrical*" interaction appears to be less intense than the "*magnetic*" one. The interaction can then be explained as follows. The *magnetic* action is responsible for the apparently more intense interaction through the κ_H "*elements*" of the two electrons. Let us suppose that the *SEA* mechanism of an electron occurs through the emission and absorption of a "*messenger*" typical of all electrons. The "*messenger*" is represented by a "*lame*" volley of the type in Fig. 13.9a (while the symmetric argument applies to positrons, which have a "*messenger*" of the type in Fig. 13.9b). But to clarify the concept better, let us consider some specific concrete cases.

The magnetic "field" of an electric current

Let's start by reporting Fig. 5.3, here redefined as Fig. 14.2. The current that flows linearly in the wire (Fig. 14.2) is almost entirely due to the electrons, because protons and ions are more or less strongly bound to the crystalline structure of the wire and in any case they have a mobility much lower than that of electrons (note that when the wire gets hot, it melts because the crystalline bonds of the atoms and positive ions break).

We point out that the present argument refers to an electron fixed in space. Obviously, if we consider many electrons located close to each other along a line parallel to the flow of current, we can say that the "field" $\boldsymbol{B}$ also moves like the current. However, this is just an inappropriate way of putting the concept. We emphasize that our axiomatic formulation rejects the same concept of "force" and "field". Therefore, in our logical scheme, it makes no sense to talk about "field movement". In other words, our formulation is different from that of concepts which, otherwise, can also be useful but only in engineering applications.

A very high number of electrons flow inside the wire, while every electron emits "*messengers*" of the type in Fig. 13.9a. Additionally, the associated signal released by every moving electron in the wire is evenly distributed across an expanding sphere.

Please note that this is not a math or "statistics" problem. The concept of 3D space is only a tool - an appropriate working scheme - a "shelf" freely chosen by the observer to organize observations (see Chapter 3). In this same sense, consider, in sharp contrast with the (erroneous) statement - generally referring to Einstein's general relativity - regarding the physical nature of the "curvature" of space. Indeed, the choice of a Euclidean space - or a curved space - is only arbitrary, devoid of any intrinsic physical meaning.

In this same respect, the isotropy of space is strictly a matter of presumption, since we have absolutely no way of distinguishing asymmetries in the path of light rays (see Chapter 3).

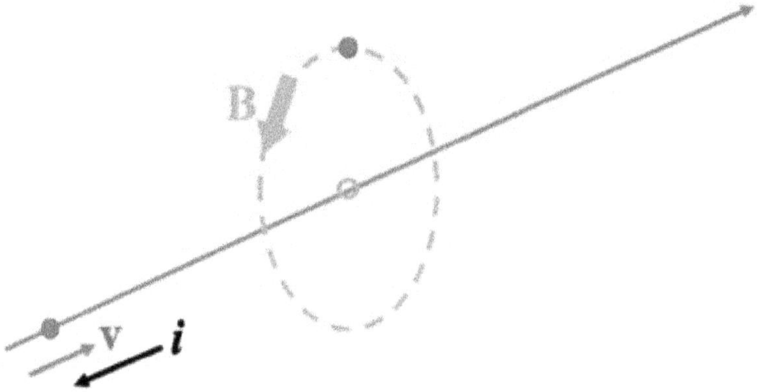

Fig. 14.2. A fixed electron, or even the needle of a magnet, represented by a small ball at the top of the figure, measures a magnetic "force" ***B*** (large arrow) generated by an electric current ***i*** shown by the black arrow: The electric current is the same as a flow of numerous electrons with velocity **v** (arrow opposite to ***i***) flowing inside an electrically conductive cable. The direction of ***B*** reverses when you reverse the direction of the electric current. The intensity of ***B*** decreases with $1/r$ where r is the distance of the fixed electron from the current-carrying cable.

The *SEA* implies that, for every single emitting electron, there is a decay in 3D space of the density of emitted "*messengers*" of the type of Fig. 13.9a, according to a $1/r^2$ law (where r is the radial distance from the source). This is a fact due to the property of Euclidean elementary geometry. It is an indisputable fact that the surface of a sphere of radius r is given by $4\pi r^2$, and the number of "*messengers*" cannot change. Their total number is always the same, and is distributed over an increasingly larger surface area.

If you integrate (i.e., you add) the effect of all the electrons flowing along the linear extension of the electric wire, you obtain - with one of the simplest formal calculations of integral calculus - a $1/r$ law, where r is the distance of the fixed electron perpendicular to the linear wire. This is in fact the correct dependence on r of the magnetic "field" generated by a current, observed from the classic measurements originally carried out by Ampère (see Chapter 5).

Furthermore, consider that all electrons flowing in the wires always have their respective "element" κ_H blocked and perpendicular to the speed of movement, i.e., perpendicular to the wire. This requirement is imposed by "*messengers*" that we do not see, but which are released by all the electrons involved in this conceptual experiment.

Suppose a fixed electron is at a given point at a radial distance r perpendicular to the linear wire, and this is essentially the same argument as in Fig. 14.2. The "element" κ_H of the "*lame*" volley of the fixed electron must twist to adapt to the same orientation of the "*messengers*" of the type in Fig. 13.9a, which are transmitted by the enormous number of electrons flowing in the wire.

This is the explanation of the magnetic field originating - according to the classical physics of secondary school and university books - from an electric current element flowing in a wire (Fig. 14.2). This holds true in the world α, while a fully symmetric argument applies in the world γ, where time and the magnetic field have the opposite direction, while the "*messenger*" is represented by Fig. 13.9b.

When the electron (in a fixed position in Fig. 14.2) moves (as in Fig. 14.3) in an environment characterized by a magnetic "field" $\boldsymbol{B}$ – which previously originated as $\boldsymbol{B}$ in Fig. 14.2 – the electron can move freely with no disturbances as far as the component is concerned of the velocity that is parallel to the $\boldsymbol{B}$ that exists in the environment. This holds as long as the "fixed" electron is at a suitable distance from the axis of the magnetic "field" $\boldsymbol{B}$ of the environment in which the electron moves. Conversely, the perpendicular velocity component causes a circular motion. The result is therefore a spiral, helical movement.

Note that this is the explanation of the law of electromagnetic induction discovered by Faraday (see Chapter 5), i.e., if a loop of electrically conductive material is moved in an environment where the "field" $\boldsymbol{B}$ varies from site to site, an electric current is generated in the loop whose purpose - as demonstrated in the classical physics of electromagnetism - is precisely to reduce or "cancel" the $\boldsymbol{B}$ value of the loop's environment.

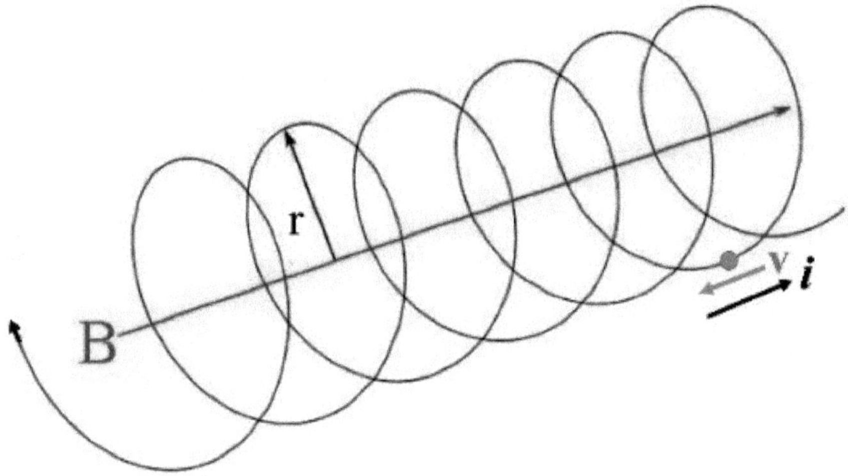

Fig. 14.3. Motion of an electron in an ambient magnetic field.

With careful reflection, we can see how this principle is the basis of all the "dynamos" that generate magnetic "fields", and of all the electric motors - whose practical development has also required studies of generations of astute and enlightened scholars. The same law applies when referring to the antitime in the world γ, where however the spiral has opposite parity.

In summary, whether this interpretation is correct or incorrect, we are faced with an important new entry into the realm of physics, that is, "chirality". That is, all natural phenomena refer to a spiral pattern, through the interaction of κ_H "*elements*" of a state – related to a huge number of volleys of moving particles. That is, chirality - illustrated symbolically by Fig. 6.3 - is an intrinsic, "endemic" characteristic of natural phenomena in the subatomic domain. See Chapter 26.

Ultimately, the key role should be highlighted that is played by the κ_H "*elements*" which are constantly released via *SEA* and mutually captured by the two interacting electrons. This crucial topic is discussed in Chapter 15 to explain gravitation - and also the controversy between Bohr and Schrödinger concerning the interpretation of the structure of the atom, or even the mysterious "*glue*" within atomic nuclei.

In Chapter 26 this same concept will be applied to the self-collimation of the solar and stellar wind, to astrophysical jets and alignments, to the shape of galaxies, and to the gravitational effect in the interaction of photons or antimatter, etc. But a suitable illustration accompanied by concretely observed images would require a large number of additional pages, which are not proposed here at the moment.

We have arbitrarily assumed that our α world is characterized by counterclockwise spirals (left-handed chirality) of miniphoton volleys and electrons that are in motion. For protons, or positive ions, the electric charge is opposite to that of the electron. In any case, due to the much greater mobility of electrons compared to protons and ions, the predominant role is that of electrons, and the "chirality" is to be expected to be mainly counterclockwise (left-handed). In principle, however, the "chirality" is the same for electrons and protons (or neutrons) while they both participate in the α-world. In contrast, the opposite chirality in the clockwise direction (right) holds for antiparticles moving in the γ-world.

Furthermore, note that, due to the equations of dynamics (see Chapter 20 and Chapter 21), an electron can eventually acquire a particularly large "*booster*", and this can imply a "kinetic" energy corresponding to a velocity greater than the velocity of light. In this case, the electron cannot retain the added "*booster*", as it can never move faster than the speed of light anyway.

Therefore, the electron must release or lose a photon - which can sometimes also materialize in a particle-antiparticle pair. This is the classic phenomenon known as Bremsstrahlung, a term now used in all languages and taken from German, whose meaning is "braking radiation".

But the surprises are not over…

Chapter 15

Atoms, Nuclei, Gravitation

"…It is not in the stars to hold our destiny but in ourselves…"
[*Julius Cæsar*, Act. I, Cassius to Brutus]
William Shakespeare (1564-1616)

…………………………………………

Scattering between photons and particles or antiparticles

Let us consider the case of a "*undegraded*" monad - i.e., a photon – that interacts with a "*degraded*" monad, i.e., with a particle or antiparticle. This is a classic of light-matter interaction, which includes three different phenomena, generally and historically referred to with different terms.

- At relatively low energy, this is the *photoelectric effect*.
- At intermediate energies, this is *Thomson scattering* and *Compton scattering*.
- At high energy the "*maximum emp density principle*" is violated by one or both monads. Therefore, when a "*degraded*" monad violates this principle, there is an emission of *Bremsstrahlung* radiation (see Chapter 14). On the other hand, when it comes to an "*undegraded*" monad, the production of particle/antiparticle pairs is observed.

Let's remember some definitions.

The photoelectric effect consists of the emission of electrons from a metallic surface when the photon has a frequency not lower than a certain threshold characteristic of every metal. The correct interpretation was given by Einstein, who was awarded for this with the Nobel Prize for Physics in 1921.

The term scattering has a broad meaning, as objects that collide may very well change their nature (as if two billiard balls could break apart and recombine into balls of different sizes).

Thomson scattering is an interaction - which appears to be almost of an elastic type - between electromagnetic radiation and a free particle with a charge. The incident photon accelerates the particle, causing the emission of radiation, with the same frequency as the incident wave. Therefore, in this way, the incident wave is diffused. It is an important phenomenon in plasma physics and was initially explained by the British physicist Sir Joseph John Thomson (1856-1940) known for discovering the electron in 1897.

Compton scattering (or Compton effect) is the phenomenon that occurs between an electron and a photon. It was observed for the first time by Arthur Holly Compton (1892-1962), an American physicist, who was awarded the Nobel Prize for Physics in 1927. The Compton effect soon became one of the decisive experimental results in favor of the quantum description of electromagnetic radiation.

A plausible mechanism for these phenomena is that the two helices of the photon interact with the "*lame*" helix of the "*degraded*" emp (i.e., the electron) with results observed experimentally in different forms depending on the used instrumental set.

Atoms

Let us consider the mechanism associated with Fig. 14.3, reproduced here as Fig. 15.1, and suppose we replace the flow of the "field" ***B*** (shown as an orange straight line) with a fixed, approximately point-like, object, which is a source of "field" ***B***, such as, e.g., an atomic nucleus that – as we well know - has its own magnetic moment. Thus, the electron *spirals around the atomic nucleus, traveling in an orbit at a fixed radial distance r from the nucleus.*

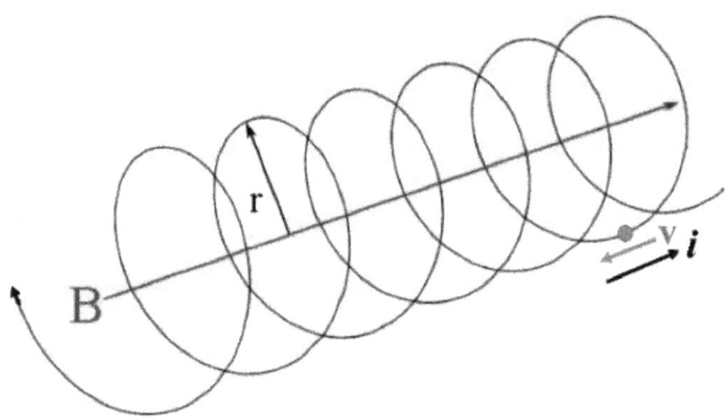

Fig. 15.1 Motion of an electron in an ambient magnetic field.

However, it remains to be explained why, in this case, the electron does not precipitate on the central nucleus. This mystery is part of phenomena internal to the atom monad. In fact, we can reasonably assume that the "elements" κ_H of the "*lame*" volleys both of electrons and of the protons found within the atomic nucleus, are such as to orient themselves appropriately and counteract the attraction due to the different electric charge.

This assumption is reasonable, although it certainly seems *ad hoc*. On the other hand, it must be considered that the spherical "shell" in which the orbit of the electron is wrapped does not have just any radius. In fact, the photon released by a "slingshot process" must satisfy a well-defined structure in accordance with what has been observed in the "spectrum" of that atom monad – a spectrum that in fact we know through the frequency of the photons that are expelled by the "slingshot effect".

That is, the "spectrum" is explained by the Schrödinger equation which – in fact – is formulated starting from a mixture of concepts associated with symbols of differential calculus (by means of continuous functions) vaguely connected to the mechanisms of Newtonian dynamics (such as energy, momentum, etc.). That is, the Schrödinger equation is an observational evidence that in some way justifies the "statistical" treatment of the quantum phenomenon within the atom monad, which we can only observe from outside.

In any case – despite all questions connected to the actual processes that take place within the atom monad – the hypothesis and argument mentioned above substantially clarify the debate, hitherto never resolved, between the conception of the atom monad according to Niels Henrik David Bohr (1885-1962) – Danish physicist, Nobel Prize for Physics in 1922 and head of the so-called Copenhagen school – and Schrödinger. According to Bohr, the electrons that orbit around the atomic nucleus should actually be considered as small, practically point-like, objects that rotate around an equally roughly point-like nucleus. That is, the model resembles a miniature Solar System. However, Bohr obviously cannot explain why the electrons do not fall into the nucleus.

Conversely, according to Schrödinger, every electron should be considered as a "cloud" of probability, consistent with his "statistical" interpretation of quantum mechanics.

Let us also remember, dutifully, that the concepts we have mentioned above are only a very small part – a very brief and very simplified final summary – of a long and very detailed learned discussion between eminent scientists.

According to our model, we can say that the interaction between the "*elements*" κ_H of the volleys – of the electron and of every proton in the nucleus – are such as to oppose the attraction due to the different electric charges of the electron and the nucleus. Therefore, every electron is unable to precipitate onto the atomic nucleus due to the interaction between the κ_H "*elements*" of the volleys. The electron can only move on a spherical surface at a suitable distance from the nucleus, in an exactly analogous way to what happens in Fig. 15.1, where the electron must move only along a spiral trajectory of constant radius.

This certainly does not seem like a complete and final answer to the problem, but it still represents an enormous step forward, even taking into account the inevitable limitations imposed by the empirical constraint. In any case, for well over a century, this problem has represented a well-known mystery, a paradox, on which rivers of ink and the efforts of the best minds have been spent. The explanation given here is certainly the only – or probably the most complete – ever given.

Nuclei

Let's consider an atomic nucleus that we know is composed of a certain number Z of protons (*atomic number* of the chemical element considered) and a suitable number N of neutrons, which would seem mysteriously necessary to keep the nucleus compact, thus preventing the protons from repelling one another because of their positive electrical charge. That is, neutrons seem to be a mysterious essential "*glue*" that holds the nucleus together.

Every chemical element is thus characterized by an *atomic number* Z (number of protons), while the *number of neutrons* N varies within certain limits, thus giving rise to various "versions" of that chemical element. Every "version" is called an "*isotope*" or "*nuclide*". The sum $Z+N=A$ is called *mass number*. Fig. 15.2 shows the complexity of the observed evidence. Isotopes can be stable or unstable, and when they are unstable, they produce radioactivity.

Let's consider the opposing action between the repulsion due to the κ_E "elements" of the volleys of every proton, and the "*glue*" due to the attraction between the κ_H "*elements*" of the volleys of every proton and also of every neutron. Evidently, the attraction between the "*elements*" κ_H of the volleys of all the protons is not sufficient to ensure the stability of the isotope. Neutrons are therefore necessary and contribute to the "*glue*" of the κ_H "*elements*" - while they do not contribute to the repulsive electric forces. Furthermore, the number of neutrons cannot be excessive, since the isotope evidently loses stability, given that the forces that determine the "*glue*" prevail to such an extent that they disrupt the stability of the nucleus.

Given the mystery represented by Fig. 15.2, this result is certainly surprising. Another paradox has thus been "unveiled", that of the stability of atomic nuclei - as well as that of the various associated isotopes.

It is an unbelievable result, considering the rivers of ink spent in the past or even today... Our journey, despite all possible warnings, is proving to be increasingly full of surprises!

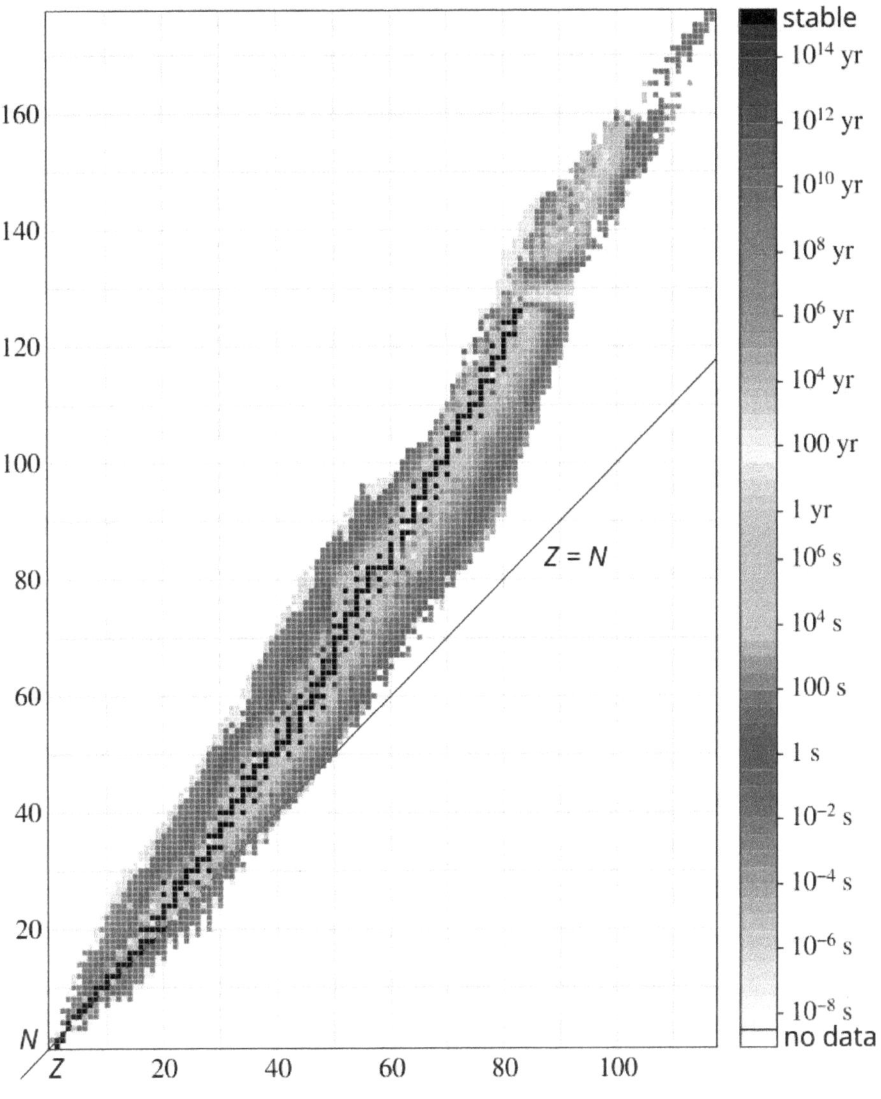

Fig. 15.2. Plot of atomic isotopes colored by half life. The darker more stable isotope region departs from the line of protons (Z) = neutrons (N), as the element number Z becomes larger (database version 6/1/2012). We see that elements with more than 20 protons, to be stable, must have a greater number N of neutrons than Z of protons. The stability of every isotope, therefore, seems to always require a number of neutrons N greater than the number Z of protons contained in the nucleus. When the neutrons are just over or just under the number required for the stability of the isotope, we have unstable and therefore radioactive isotopes. All this is a well-established observed fact, and is generally considered mysterious, unexplained, a true paradox. After *Wikipedia, List of elements by stability of isotopes.*

Gravitation

Finally, let's consider the paradoxes associated with universal gravitation, which is afflicted by the "original sin" represented by having taken action at a distance, with instantaneous propagation of the gravitational signal (see Chapter 5). In fact, this "original sin" is still perceived as a serious handicap in the most advanced theoretical physics works. The gravitational "field" is said to be a case that is "elusive" to any attempt at understanding and "quantization" – such as was done for the electromagnetic "field". The gravitational "field" seems in many ways paradoxical and mysterious.

In order to resolve this mystery, some marvelous, very delicate experiments are being carried out by some of the most learned and authoritative scientists, to try to measure sudden and short-term variations in the gravitational signal arriving from space. Man has never stopped exploring the mysteries of Nature!

Concerning our conceptual journey, we refer to Ockham's razor (see Chapter 14) and consider whether it is really necessary to suppose the existence of an interaction with miniquanta - which are similar to miniphotons but associated with gravitational interaction and of a different nature from miniphotons which are associated with electromagnetic interaction. In fact, we will see that the miniphotons already defined seem to be well sufficient to explain the gravitational interaction as well.

In short, by relying on the language of Newton's "force," the principle is simple. In general, it is believed that the force of *attraction* between two unit electric charges of opposite sign is *identical*, with a changed sign, to the force of *repulsion* between two unit electric charges of the same sign. This is consistent with all laboratory measurements, but only within a certain range of experimental error. In fact, this equality between attraction and repulsion may not even be exact, but it could differ by a very small amount measurable only when a truly enormous number of electric charges are considered. Let's see in detail how this topic is expressed in the language of "*lame*" spirals.

Newton invented universal gravitation on a purely speculative basis, just as an *ad hoc* working hypothesis. In fact, as already mentioned in Chapter 4, Kepler had obstinately investigated whether all previously known forces (mainly magnetism) were suitable for explaining the empirical laws that he had marvelously found in the motion of the planets. Kepler's repeated attempts were all frustrating. Indeed, it was not possible to avoid the classic *ad hoc* hypothesis formulated by Newton.

According to our journey, we consider two "*degraded*" monads, with masses M_1 and M_2, respectively, each with a strictly neutral electric charge.

That is, we are referring to two monads, both located in the world α or both in the world γ, and each monad is the set of a very large number of smaller monads, as is precisely the case for the Sun or for a planet or planetary object that contains a huge number of atoms, ions, nuclei, electrons, etc. Newton's universal gravitation hypothesis states (see Chapter 4) that the two monads attract each other with a force $\boldsymbol{F} = (G\ M_1 M_2)/r^2$ where r is the distance along the line connecting the two monads. But, in our approach, we must reject the concept of "force".

The two monads can only communicate through the *SEA* mechanism (see Chapter 14) thanks to their respective mini-monads (or "*virtues*"), emitted either by the "*slingshot effect*" - or through electromagnetic signals that are released almost like a "*rubber band*", i.e., "*double helix*". Since both monads have strictly neutral electrical charge, we can envisage two possibilities.

In one case all the respective "*virtues*" do not have pairs κ_E and k_H that are decoupled. However, in this case the monads are photons.

The other possibility is that each monad - which is electrically neutral - has exactly the same number of smaller monads with positive and negative electrical charge, so that, statistically, the number of miniphotons - which have unpaired pairs of an electrical charge - is exactly equal to the number of miniphotons - with unpaired pairs of opposite electric charge.

We stress that, due to the isotropy of space and time (see Chapter 3, Chapter 20 and Chapter 21), it does not make sense to speculate on a possible deviation - compared to a perfect spherical symmetry - of the gravitational interaction, because it is just impossible to detect any anisotropy. In fact, there is no reference system for comparison, i.e., an alternative system to photons. - and, owing to this assumption, we can evaluate no deviation from isotropy in 3D space.

The monad with mass M_1 is composed of a number N_{1+} of positively charged minimonads, plus a number N_{1-} of negatively charged minimonads, where $N_{1+} = N_{1-} = N_1$. Define the analogous quantities referring to M_2. Therefore, the electromagnetic interaction through miniphotons is as follows

repulsion	$N_{1+} \cdot N_{2+} \cdot \varepsilon_+ + N_{1-} \cdot N_{2-} \cdot \varepsilon_- = N_1 \cdot N_2 \cdot (\varepsilon_+ + \varepsilon_-)$
attraction	$N_{1+} \cdot N_{2-} \cdot \varepsilon_0 + N_{1-} \cdot N_{2+} \cdot \varepsilon_0 = 2 \cdot N_1 \cdot N_2 \cdot \varepsilon_0$
total balance	$N_1 \cdot N_2 \cdot [2\varepsilon_0 - (\varepsilon_+ + \varepsilon_-)]$

where ε_+ is the total content of emp exchanged in a *repulsion* between positive electric charges, ε_- is the total content of emp exchanged in a *repulsion* between negative electric charges, and ε_0 is the total content of emp exchanged in an attraction between various electric charges.

For the reader less familiar with formulas, we stress that the calculation above is identical to what is done when drawing up a family budget between income and expenses.

Thus, whenever $[2\varepsilon_0 - (\varepsilon_+ + \varepsilon_-)] > 0$, gravitation can be explained by the electromagnetic "*messenger*", with no need to appeal to any new type of interaction.

Notice the proportionality $N_1 \cdot N_2 \propto M_1 \cdot M_2$ (where the symbol $\propto$ denotes "proportional") and this is consistent with Newton's universal gravitation. That is, the differences between ε_+, ε_- and ε_0 are so tiny that we can only observe them when we consider bodies with very large masses M_1 and M_2, one of which has at least the size of a planet or a body in the Solar System. Moreover, note that, owing to the same argument applied to Coulomb's law (see Chapter 5), the result is also strictly consistent with the $1/r^2$ dependence of Newton's universal gravitation.

Finally, let us consider the case of two "*degraded*" monads, respectively with masses M_1 and M_2, characterized by a *rotational* emp content. If M_1 and M_2 are expressed in terms of the total emp content, one can ask whether the rotational energy (i.e., the associated emp content) also enters the gravitational interaction. Note that "spin" is also part of the rotational energy balance, although "spin" is just a "*quantum number*" - while, due to Ockham's razor, we want and we need to avoid "hidden parameters" that imply some model or arbitrary hypothesis, at least as long as no further observation is available from the breaking up of monad particles.

In principle, the interaction between the smaller positive and/or negative monads of every monad is certainly affected by their relative velocity. However, by considering the "statistical" effect resulting from the sum of the smallest monads, the contribution of the rotational emp could be statistically canceled out. This is, at least, a reasonable hypothesis. This property, however, has been studied as briefly reported in Chapter 20 and Chapter 21. This also applies to the role of "kinetic" energy and its possible influence on gravitation.

In other words, when we consider an object with *rotational* emp content, we refer to an object composed of several, or very often numerous, monads. Therefore, all of this is, in any case, part of a "statistical" treatment. The interaction between single monads uses a quantum number (or "*virtue*") called "spin". In fact, even a monad that we consider "as small as possible" has its own internal structure that we ignore due to the empirical constraint. The "spin" is defined in order to overcome our inability to access the internal mechanisms of the monad.

In any case, in principle, the final total gravitational attraction of the two monads should also depend on their respective rotational emp content. Ultimately, the concept of rotational energy is a "simple" and convenient algorithm conceived by the human mind to carry out calculations in terms of simplified "statistical" analysis, or "thermodynamic" analysis – and this is very useful for applied science – while in itself it can have no real basic axiomatic physical meaning.

In conclusion, gravitation is *not* a new type of interaction, in addition to electromagnetic interaction - even if our assertion, according to today's most advanced theoretical physics, constitutes "heresy".

We also emphasize that, according to the mechanism considered in our logical journey, two photons - having no "*lame*" monads - can have no mutual gravitational attraction or repulsion. On the contrary, a ray of light immersed in an environment dominated by "*messengers*" emitted by a body of zero electrical charge, but which contains elements with "*lame*" volleys, is certainly affected by the gravitational action of the body of zero electric charge. The surprises never seem to end!

The reader with a little background knowledge will certainly have read about a famous astronomical experiment. The position of visible stars – for a short time lag during a total solar eclipse - clearly showed how the light rays of a star are significantly deflected by the gravitation of the Sun. This experiment aroused great interest, and it was said - and is said still – that this proves that space is "curved", consistently with the model of general relativity which is based on the assumption that space is curved - and where the classical Euclidean geometry must therefore be replaced by a new geometry to be treated with a very complicated formalism of mathematics (tensor calculus). This phenomenon has been and is repeatedly observed, and is referred to as "gravitational lensing".

According to our logical journey, this statement must be completely revised. In fact, due to the *SEA* mechanism (see Chapter 14), when a photon is found in a region of space with a rich presence of "*messengers*" originated by a sufficiently large object in terms of emp content – i.e., by an object of sufficiently large mass – the photon interacts with these "*messengers*" and necessarily must deviate. The effect is exactly as if space had been "curved" by the gravitational field of the object.

In other words, the assertion - that the choice of Euclidean space or curved space is only an arbitrary choice of the observer, only aimed at providing a more or less intuitively simple descriptive model - remains valid in all respects. But there is nothing physical in this choice, which is only a decision of convenience for an (illusorily?) "simpler" description. We will return to this same topic in Chapter 20 and Chapter 21.

Before proceeding further, it is now necessary to "*wash our dirty laundry a little*" in order to free ourselves from erroneous conceptions that are today universally (but not too much...) reported and considered as well established, and accepted by the entire scientific community. Indeed, the surprises on our journey never end...

Chapter 16

Einstein's Mistake

"There are more things in heaven and earth, Horatio,
Than are dreamt of in your philosophy."
(*Hamlet*, Act I, Scene, IV, 166-167)

....

Axioms and clock synchronism

Einstein introduces two axioms, i.e., assumptions considered true, with no possibility of being called into question.

A first axiom is "*realism*" (see Chapter 3), which is defined by A. Einstein, B. Podolsky, and N. Rosen, 1935 as follows: "*any serious consideration of a physical theory must take into account the distinction between the objective reality, which is independent of any theory, and the physical concept with which the theory operates.*"

Einstein (1955; in his chapter on "*The evolution of the concept of natural reality*") claims that the "*faith*" in this axiom is the basis of all natural science. In other words, the *principle of "realism"* maintains that the occurrence of a phenomenon is independent of the observer who observes it.

Einstein's second axiom states that nothing can travel faster than the speed of light.

Einstein, in his historical memoir of 1905 - perhaps the most famous - makes the following argument, which is now called the "*synchronism*" argument. Given its crucial importance, we report the argument in detail here. The required mathematics is very little - and intuitively understandable - and every reader should follow it all easily and patiently.

The initial concern deals with what Einstein calls "*definition of simultaneity*". Einstein's concern is about the *synchronization* of clocks of two different observers A and B, who are stationary - i.e., not moving with respect to each other. The "*time according to A*" is, e.g., provided by the position of the hands of a clock in the possession of A. "*Time according to B*" is similarly provided by the position of the hands of a clock that is ideally identical to the previous one and is in the possession of B.

A ray of light is emitted from A at the instant of "time of A" called t_A and is reflected by B at the instant of "time of A" called t_B. The reflected ray of light arrives at A at the instant of "time of A" called t'_A. Therefore, "*in agreement with the definition of synchronism of the two clocks of A and B*" we have

$$t_B - t_A = t'_A - t_B \qquad (16.1)$$

Furthermore, consistently with experimental measurements of the speed of light c, we have

$$\frac{2AB}{t'_A - t_A} = c \qquad (16.2)$$

where AB denotes the length of the linear segment between A and B, and $t'_A - t_A$ is the time required by the ray of light to complete the path AB in both directions, outward and return. Up to this point there is nothing to complain about.

However, some doubts concern the following paragraph which begins verbatim as follows.

"*The following reflections are based on the principle of relativity and on the principle of the constancy of the speed of light. These two principles we define as follows:*

1) The laws by which the states of physical systems undergo change are not affected, whether these changes of state be referred to the one or the other of two systems of coordinates in uniform translatory motion.

2) Any ray of light moves in the "stationary" system of coordinates with the determined speed c, whether the ray be emitted by a stationary or by a moving object. Hence

$$velocity = \frac{light\ path}{time\ interval} \qquad (16.3)$$

That is, these two axioms are, respectively, that of "*realism*" (here called "*principle of relativity*") and that of constancy of the speed of light. For what follows, note that (16.3) can also be rewritten as follows (with concepts familiar from the elementary algebra taught in middle schools)

$$time\ interval = \frac{path\ travelled\ by\ the\ ray}{velocity} \qquad (16.4)$$

Now Einstein imagines an ideal experiment, with two systems of reference, where one system of an observer A is said to be "*stationary*", while the second is that of an observer B who moves, with respect to the stationary system of A, with speed $v = |\mathbf{v}|$ constant and uniform, directed along the axis $\boldsymbol{x}$.

Einstein considers an ideal stationary *rigid bar* of length l which is measured with another *sample bar* which is also stationary. Einstein imagines that the rigid bar is aligned to the $\boldsymbol{x}$ axis of the "*stationary*" reference system and that it is placed in the reference system moving uniformly along the same $\boldsymbol{x}$

axis with speed **v**. See Fig. 16.1 below. The standard bar is also placed, identically to the main bar, in the reference system in uniform motion.

Einstein now addresses the problem of measuring the length of the bar, which is placed in the reference system of B, by the observer A who is in the "*stationary*" system. The stationary observer A sends light rays to the two ends of the bar and measures the return time of the rays reflected from the two ends of the bar. Everything is always referred to the time measured by the "stationary" observer A. Einstein also imagines that two further observers are placed on the two ends of the rigid bar, which is in motion with respect to the "stationary" system.

Einstein claims (*verbatim*) that, applying the same criterion for the measurement of the synchronism associated with (16.1) and (16.2), one finds

$$t_B - t_A = \frac{r_{AB}}{c - \mathrm{v}} \tag{16.5a}$$

$$t'_A - t_B = \frac{r_{AB}}{c + \mathrm{v}} \tag{16.5b}$$

where r_{AB} is the length of the rigid bar measured in the "stationay" system.

Note that if the rigid bar is reduced to a point, and if the bar is not in motion but is stationary, we have the case expressed by (16.1).

In the two formulas (16.5) the two denominators $(c - \mathrm{v}\,)$ and $(c + \mathrm{v})$ denote the fact that the space traveled by the ray of light is different between the "stationary" observer and the extremes A and B of the bar (don't get confused from the fact that the two observers are called A and B, as well as the two ends of the bar, the coincidence has no meaning; it would be better to call, e.g., P and Q the two ends of the bar for example) and this space is different at different times due to the motion of the bar.

There is no need to go into detail on this argument which may also seem rather nebulous. However, the argument is more or less intricately connected to the measurements of instants of time expressed in the "stationary" system of the observer A – i.e., the instants used in the two formulas (16.5). Each formula of (16.5) is of the type (16.4), i.e., it considers that the instants - in which the reflection takes place of the light signal sent by the "stationary" observer A - change due to the displacement of the rigid bar which is moving with speed **v**.

That is, in the two relationships (16.5) the speed of light is expressed as $(c - \mathrm{v})$ and $(c + \mathrm{v})$. In other words, in reality, this implicitly hypothesizes that the constancy of the speed of light no longer holds, but is added or subtracted with **v**.

This is logically possible, as long as it is taken into account. However, Einstein comments: "*so we see that we cannot attach any* absolute *signification to the concept of simultaneity, but that two events which, viewed from a system of coordinates, are simultaneous, can no longer be looked upon as simultaneous events envisaged from a system which is in motion relatively to that system.*" This conclusion is incorrect!

In fact, let's consider the general case of assuming two hypotheses that we will call H1 ("*realism*") and H2 (constancy of the speed c of light). We drop H2, and we obtain a result which is not consistent with the hypothesis H1. This means that H2 is *a necessary condition* for H1 to be true. But certainly, this fact absolutely *does not* mean that "*we have demonstrated*" that H1 is not valid!

On the objective necessity of considering simultaneity

We can now ask ourselves the question whether "*simultaneity*" is a property to be considered implicit in "*realism*".

If "time" is absolute - as the Greek philosophers hypothesized, and as Einstein certainly hypothesized (see Chapter 9) – simultaneity is an intrinsic part of "*realism*", and if simultaneity is different for two observers, "*realism*" is certainly violated.

If we also consider two observers, one in the world α and the other in the world γ, obviously they observe different phenomena, but one performs observations with respect to time, the other with respect to antitime. Therefore, simultaneity has a necessarily distinct meaning in the world α and world γ. It makes no sense to talk about the simultaneity of phenomena that occur in spaces where "time" is defined in a totally different way than antitime. In any case, moving an observer with uniform motion with respect to another is certainly not equivalent to moving an observer from world α to world γ or vice versa.

"Time" is one and identical throughout the world α, and antitime is one and identical throughout the world γ.

In conclusion, if we write (16.5), we might as well assume directly - without any preamble - that events that are simultaneous in one reference system are not simultaneous in another. However, this is evidently an *a priori* violation of the claimed axiom of "*realism*" – and is in manifest contrast with the "*faith*" claimed by Einstein above.

In any case, Einstein built the entire remaining part of his "historical" memory of 1905 by relying on this misunderstanding. Furthermore, in the substantial formal mathematical construction of his historical memory, Einstein repeatedly uses expressions of the type $(c - \mathrm{v})$ and $(c + \mathrm{v})$, i.e., he repeatedly violates the axiom of constancy of the speed of light. There is no doubt about this.

But, regardless of all this, if "*realism*" holds - and if "time" and "antitime" characterize the world α and the world γ, respectively - for what purpose should we worry about the "*synchronism*" of two clocks used by different observers?

Every observer - as well as every living being - has (in the world α) the perception of "time", and this perception is independent of the clock used by either one or another observer. "Time" is unique in the α world and is something that meets the requirements of "*realism*". Similar considerations apply in the world γ with reference to the antitime.

Furthermore, it is reasonable to ask the question of the meaning of "rigid bar". A "rigid" body is an abstraction - since nothing in Nature is "rigid" at all. Everything deforms, even just due to small efforts or small variations in temperature. Furthermore, rigid objects do not exist in a star. In the Universe only rare exceptions are represented by sufficiently cold environments (such as the Earth) where the solid state can exist. Furthermore, Einstein considered a rigid bar to be an immaterial object, a mere geometric abstraction. However, one must ask: *is it necessary indeed to consider such an immaterial object?*

It should also be considered that all of Einstein's derivation is based on the principles of dynamics according to Newton, with the concepts of force, field, mass, etc. which are excluded from the formulation adopted in our journey. In other words, all this must necessarily be substantially rethought critically.

Consequences of the mistake

This Einstein "mistake" had very important (and devastating) consequences. Among other consequences, Einstein demonstrates the contraction of lengths, defined as a reduction in the measurement of the size of an object, but *only dealing with the direction of its uniform rectilinear motion* with respect to a "stationary" observer A (Fig. 16.1). In fact, the formula for measuring the length of an object in motion with respect to a "stationary" observer is shown by Einstein to be

$$L = L_0 \sqrt{1 - \frac{\mathrm{v}^2}{c^2}} \qquad (16.6)$$

where L_0 is the "*proper*" length (i.e., the length measured by an observer at rest with respect to the object), L is the length measured by an observer in motion relative to the object, and $\mathrm{v} = |\mathbf{v}|$ is the relative velocity between the "stationary" observer and the object. The following factor is called Lorentz factor, after the Dutch physicist Hendrik Antoon Lorentz (1853-1928; joint winner, with Pieter Zeeman, of the Nobel Prize for Physics in 1902 for his theory of electromagnetic radiation)

$$\gamma = \frac{1}{\sqrt{1 - \frac{v^2}{c^2}}} \tag{16.7}$$

and this factor frequently occurs in Einstein's formulas. Note that the Lorentz factor is always greater than or at most equal to 1.

Note that the length contraction occurs *only* along the direction of the velocity $\mathbf{v}$, while it does not occur along axes perpendicular to $\mathbf{v}$. That is, geometric figures are substantially deformed (like in a distorting mirror). Furthermore, the dependence on $\mathbf{v}^2$ ensures that the result is independent of $\mathbf{v}$ or $-\mathbf{v}$, i.e., of the direction of movement - and this, as we will see in Chapter 20 and Chapter 21, raises serious paradoxical problems. Something is wrong.

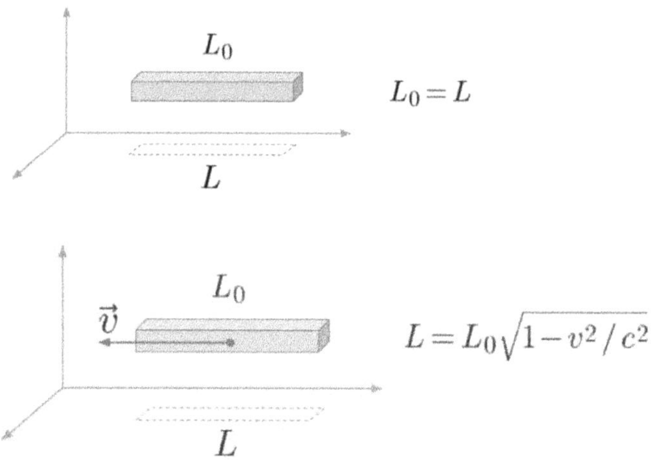

Fig. 16.1. Relativistic length contraction. See text.

Furthermore, an observer at rest looking at an object moving at the speed of light should ideally see its dimension in the direction of motion as equal to zero.

Another curious consequence is the dilation of time. That is, consider an effect for which the duration of a phenomenon - defined as a succession of events - is minimum when the time interval between the initial and final events is measured in a reference system that we call S and which is at rest with respect to the two "limit" events (initial and final). S is called the reference system *united* to the phenomenon. The duration Δt of a phenomenon - measured in a *united* reference system - is called *proper time*.

However, Einstein shows that for all reference systems moving with respect to S the duration $\Delta t'$ of the phenomenon is larger. Therefore, the duration of a phenomenon is not the same in two reference systems in relative motion. According to Einstein, the relationship that links the two different time intervals is

$$\Delta' t = \gamma \, \Delta t \tag{16.8}$$

where $\Delta t'$ is the time interval measured in a reference system S' moving relative to S, and Δt is the *proper time* interval. Furthermore, as already stressed, it is important that the coefficient γ depends on $\mathbf{v}^2$ and this fact means that the result is independent of the direction of movement, i.e., of $\mathbf{v}$ or $-\mathbf{v}$.

Another concept connected to all this is the relativistic mass m defined as the product between the rest mass m_0 of a body and the Lorentz factor

$$m = \gamma \, m_0 \tag{16.9}$$

In fact, this relationship was introduced by Lorentz in 1904 as part of his *theory of the electron* dating back to 1892, i.e., *13* years before Einstein's famous memoir of 1905. In this theory the electron was represented as a charged sphere, which underwent a contraction of length in the direction of motion (Lorentz contraction). Lorentz also wrote two equations, for the "longitudinal" and "transversal" mass of the electron, respectively. The transverse mass was equivalent to $\gamma \, m_0$ whereas the longitudinal mass corresponded to $\gamma^3 \, m_0$. Lorentz struggled desperately to get to the bottom of the intrinsic paradoxes in Maxwell's equations (see Chapter 5) resulting from the finite (i.e., not infinite) propagation speed of electromagnetic waves.

Many distinctions would now be necessary - which we will spare here - connected to the authorship of these concepts and the history of thought related to them. The fact is that Einstein deluded himself into thinking that with his argument on "synchronism" he had found the key to explaining many embarrassing "oddities" caused by the finite speed of light found within Maxwell's equations (see Chapter 5).

As a curiosity, let's also remember the so-called "twin paradox". One of the twins hypothetically goes on a journey on a spaceship, and returns after a certain time. The spaceship's speed is high, and time contracts according to (16.8). When centuries or millennia have passed on Earth, time has undergone a relativistic contraction on the spaceship, therefore, the traveling twin has aged slightly, etc. But the same argument applies, vice versa, to the twin on the spaceship who considers himself to be "stationary", and the twin on Earth as the twin in motion. So the question arises: which twin remained "young" and which one aged and passed away? Rivers of ink have been written on this paradox, which is evidently generated by the dependence of the Lorentz factor γ on $\mathbf{v}^2$, or rather this is a consequence of Einstein's mistake.

The history of the acceptance of the theory of relativity is very tangled (see Chapter 17). Einstein's mistake - certainly noticed by several scholars in the initial stages of the discussion - was later set aside. G. P. Gregori "rediscovered" it, and – with the encouragement of some authoritative scientists – published a small volume in 2005. He was subsequently invited to present the volume in September 2006 at a specialist conference at *Imperial College* (London). The reception from the approximately *50* participants was comforting ("*it's not surprising that Einstein made a mistake, it's rather surprising that for a century no one ever noticed it*").

As a matter of fact, however, the concept of relativistic mass has now been used with apparent success - and for many years - in the design of large particle accelerators, and its use appears essential. In other words, there is something true in all this formalism. We return to this item in Chapter 20 and Chapter 21 and we will show, through the discussion of the symbols $\xi(|\mathbf{v}|)$ and $\chi[(\mathrm{v}/c)]$, how the contraction/expansion of mass, i.e., the concept of relativistic mass, is strictly necessary in the formulation proposed here.

Another consequence, however, is very different - with very serious implications in celestial mechanics and astrophysics. The concern is about the Doppler Effect (see Chapter 11) when calculated with relativistic formulas - and which is fundamentally different from the Doppler Effect calculated with classical dynamics formulas. In fact, the relativistic formula of the Doppler Effect is usually used in astrophysics to measure the motion of galaxies with respect to the Earth and therefore the expansion of the Universe (big-bang etc.). In an authoritative article on astrophysics, it is written that one must hope that the formula is correct, otherwise all estimates on the expansion speed of the Universe will have to be redone.

To verify whether the relativistic formula of the Doppler Effect is correct or not, it is necessary to know an object that moves at a known and sufficiently large speed with respect to the observer - and according to a well-known trajectory - and thus verify whether the observed Doppler Effect is in agreement with the formula to be checked.

The only such experiment available today is represented by the trajectories of the two *Pioneer* space probes (Fig. 16.2). The trajectories were calculated with extreme accuracy by means of celestial mechanics, based on Newton's law of universal gravitation, a law that is certainly very valid, even though - according to the point of view illustrated here - it is valid in a "statistical" sense (see Chapter 18, Chapter 20 and Chapter 21). In fact, the innumerable successes of the space probes - launched even for encounters with tiny objects such as comets or asteroids - are irrefutable proof of the reliability of the orbits calculated by celestial mechanics for the two *Pioneer* probes.

On the other hand, the trajectories of the *Pioneers* can be computed based on the relativistic Doppler Effect of the received radio signal. This can be done either by considering the Doppler Effect associated with the orbital velocity parameters, or by considering the Doppler Effect associated with the probe spin parameters (spin is necessary for the stabilization of the probe). In both of these cases a surprising and mysterious discrepancy is found with the orbital parameters calculated by celestial mechanics.

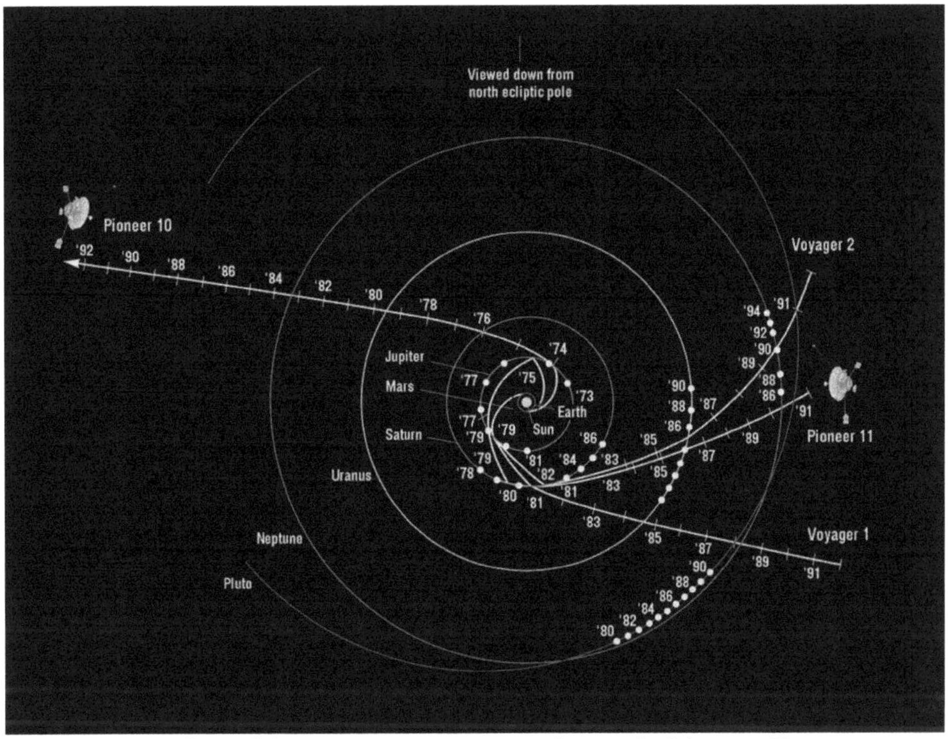

Fig. 16.2. Orbits of the two *Pioneer* space probes and of the two *Voyager* space probes. *Pioneer 10* was launched on March 2, 1972. *Pioneer 10* is currently (in 2024) 11.73 $billion\ km$ from Earth and traveling at 44788 $km/hour$ relative to the Sun. At that distance, signals take 21 $hours$ and 45 $minutes$ to arrive on Earth. The twin *Pioneer 11* probe was launched on April 6, 1974, with the last contact being on November 24, 1995. The two *Voyager* probes are less useful for discussing the Doppler Effect. See text.

Rivers of ink have been spent on this distressing paradox, studied by hundreds of the most expert scholars, and it is still being studied. This denotes that the relativistic formula used for the Doppler Effect is certainly wrong, and this experiment is the only one available today to verify the correct formula of the Doppler Effect to be used in astrophysics. Note that the orbital parameters of the two *Voyager* probes are less suitable for this purpose, and have essentially not been used for this purpose.

Conclusion

The axiomatic formulation of relativity contains, in essence, at least some paradoxical logical steps - and if we want to avoid paradoxes, we certainly cannot use *sic et simpliciter* the formulation reported in standard specialist texts on the theory of relativity, which largely adopt the concepts of the dynamics of Newton (such as mass, force, etc.). It is really necessary to proceed very cautiously, beginning from the experiments that - it is claimed - “prove” the theory of relativity. That is what we shall do in Chapter 17.

Chapter 17

Michelson and Morley – Georges Sagnac
Physicists and Philosophers
Debates and Misunderstandings

"Liber est autem, qui seruitutem suam effugit: hæc est assidua et ineluctabilis et per diem ac noctem æqualiter premens, sine interuallo, sine commeatu. Sibi seruire grauissima est seruitus: quam discutere facile est … Ad hoc proderit nobis inspicere rerum naturam …"

"… He is the true freeman who has escaped from bondage to self. That slavery is constant, from it there is no deliverance; it presses us day and night alike, without pause, without respite. To be a slave to self is the most grievous kind of slavery; yet its fetters may easily be struck off … To this end it will be profitable for us to examine the nature of the universe …"

Book III, Which treats of the different forms of water, Preface, 16-18
[translation by John Clarke]
Lucii Annæi Senecæ, *Quæstiones Naturales Liber III De æquis, Præfatio,* 16-18;
[lived (4 BC (?) - AD 65); text written in AD 62-65]

……………………………………………

Our journey becomes increasingly difficult, even though fascinating (and sometimes sad). We are talking about some experiments, often contested, which - it is claimed - prove the correctness of the theory of relativity.

Michelson & Morley

Let's start with the description of the Michelson interferometer - invented by Albert Abraham Michelson (1852-1931), who, born in Poland, later became an American physicist and the first head of the *Physics Department* at the *University of Chicago*.

See Fig. 17.1. A light source (e.g., a laser) sends a ray of light, of approximately a single wavelength, onto a semi-reflective mirror (semi-silvered, in the figure called a "*beam splitter*") which directs two rays of light in different directions. The two rays must be coherent with each other, i.e., the respective electromagnetic waves must be perfectly in phase, which means that every maximum of the wave of one ray corresponds to a maximum of the wave of the other ray, etc. The two coherent light rays are sent to two mirrors placed, respectively, in A and in B (in the figure, they are indicated as "*Mirror 1*" and "*Mirror 2*", respectively). The mirrors A and B reflect the received ray, and send it back to the semi-reflective mirror. This semi-reflective mirror sends the two rays - received back from the mirrors A and B - to the observer who records the signal, e.g., on a photo plate.

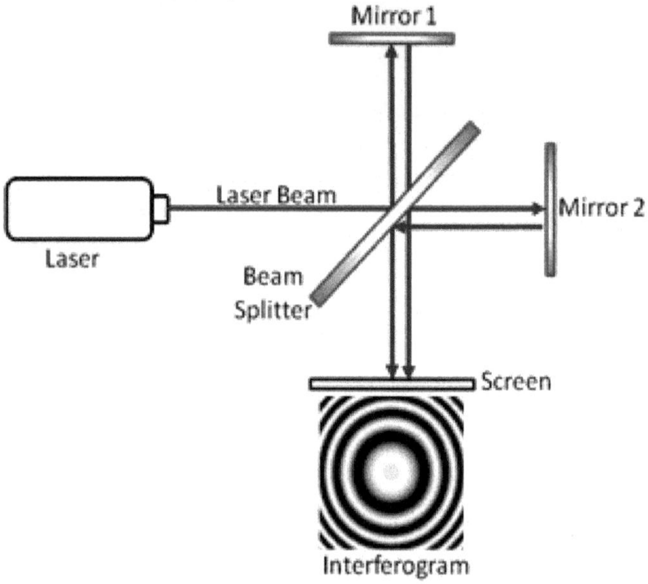

Fig. 17.1. Michelson interferometer. See text.

If one of the two distances changes between the semi-reflective mirror and the mirrors A and B, respectively, the observer sees a change in the interference fringes on his photographic plate – i.e., he sees moving luminous bands. In fact, interference fringes are a phenomenon that is a classic of geometric optics, and this is well known to all secondary school students (see Chapter 11).

Today Michelson's interferometers are in operation with arms even a few kilometers long, aimed to study the time-varying gravitational fields that arrive from the cosmos. They are marvelous experiments (*LIGO, VIRGO*, etc.) that engage the best minds in physics today.

We now describe the famous experiment by Michelson and Morley, performed for the first time in 1881, i.e., *24* years before Einstein's famous memory. Edward Williams Morley (1838-1923) was a US physicist who collaborated with Michelson.

Newtonian physics assumed that every type of wave (light, sound, etc.) must propagate through some elastic medium, e.g., sound waves propagate through air. Similarly, for electromagnetic waves, the existence of an elastic medium similar to air, called “æther”, was hypothesized. Therefore, if this hypothesis is correct, the speed of light predicted by Maxwell’s equations should be that relative to an “absolute” reference system, which is stationary with respect to the æther. Consequently, every object in motion in the Universe should be subject to the effects of its motion with respect to the æther. In other words, every object should observe a sort of “æther wind”.

Therefore, a generic observer moving with respect to the “absolute” system of æther, when measuring the speed of light in different directions, should measure different values by applying the Galilean principle of velocity composition. For instance, an observer on Earth, due to the motions of rotation and revolution of the Earth, should observe an aether wind, which can be calculated at approximately 30 km/sec.

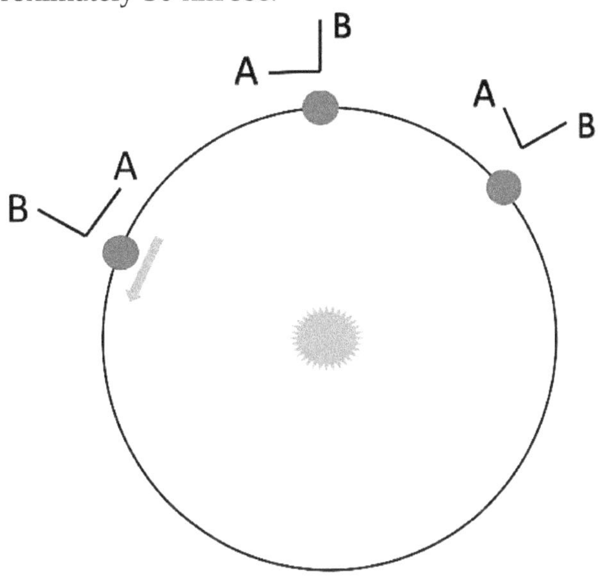

Fig. 17.2. Out of scale figure. The Earth, a small ball, rotates around the Sun, the central ball, counterclockwise when observed from the Pole Star. A Michelson interferometer can be operated with one arm parallel to the Earth’s motion, and the other arm perpendicular to it. With this configuration, you should see well-recognizable interference fringes. If, however, the orientation of the two arms of the interferometer is random and different compared to the astronomical motion of the Earth, the result is more or less marked, but far less clear. See the text.

See Fig. 17.2. The arm of the Michelson interferometer between the semi-reflecting mirror and mirror A is placed along the direction of revolution of the Earth’s motion, as it moves around the Sun. The other arm of the Michelson interferometer, towards mirror B, is perpendicular to the Earth’s motion around the Sun. Given the speed of the Earth's motion, the length of the arm of mirror A must therefore detect a speed of light that has changed due to the motion with respect to the "absolute" reference system of æther. Conversely, the speed of light measured along the arm of mirror B must be unchanged. If the arms towards A and towards B have any direction – i.e., not specifically along the Earth's orbit or perpendicular to it - the disturbance due to the "æther wind" should be less marked and partially affect both arms of the interferometer. Therefore, by exploiting the motion of the Earth it was thought that it is possible to measure the motion of the Earth with respect to the "absolute" reference system represented by æther.

Michelson actually wasn't thinking about the theory of relativity. Michelson was a renowned scientist at the time, and his main goal was the definition of unit standards (e.g., the meter standard, etc.), including the possible dependence of unit standards on Earth's motion. So he carried out his experiment for a very specific purpose. The experiment was then misunderstood, due to Einstein's aforementioned "mistake" (see Chapter 16). The experiment gave an entirely negative result.

It must be emphasized that these experiments require extreme precision (nor Michelson had a laser source that had not yet been discovered). To obtain more precise data, in 1887, Michelson got in touch with Morley, who offered him his basement for a new experiment for which an interferometer was mounted on a square stone slab 1.5 m on a side, and about 30 cm thick. We leave out other details. It was concluded that æther does not exist. Michelson was the first American to be awarded (in 1907) the Nobel Prize for physics.

In reality, his experiment had to *necessarily give a negative result*. The misunderstanding was due to the fact that the Lorentz factor [see (16.6) and (16.7)] depends on $\mathbf{v}^2$ and is therefore *independent* of the direction of $\mathbf{v}$. In fact, the Michelson and Morley's experiment in itself entails that the ray of light that runs along the direction of the Earth's motion travels the identical path once in one direction, and immediately afterwards in the other. Therefore, if the effect depends on the direction of the ray, the two effects, added back and forth, cancel each other out exactly! This aspect will be explained more fully in detail in Chapter 20 and Chapter 21, where it is shown how the Lorentz factor must be replaced by another treatment, consistent with the ABC theorem (see Chapter 3).

The substitutive factor of the Lorentz factor, however, depends on the direction of $\mathbf{v}$, i.e., it must depend on *odd* powers of $|\mathbf{v}|$, i.e., on v, v^3, v^5, v^7, ..., and *not* on even powers of $\mathbf{v}$.

Consider Figure 17.3. Suppose we use a single source to generate two coherent rays of light - to study their interference after the two rays have traveled identical paths in opposite directions. Indeed, you can never see any difference if the Lorentz factor depends on $\mathbf{v}^2$ and is independent of the direction of $\mathbf{v}$. In Fig. 17.3, a large number of reflective mirrors are indicated (they are 9, but they can be as many as you like), such as to ensure that each of the two rays of light returns again to the partially reflecting mirror after having completed the same path. Each ray will have completed the same path in the opposite direction to the other, thus giving rise to interference fringes.

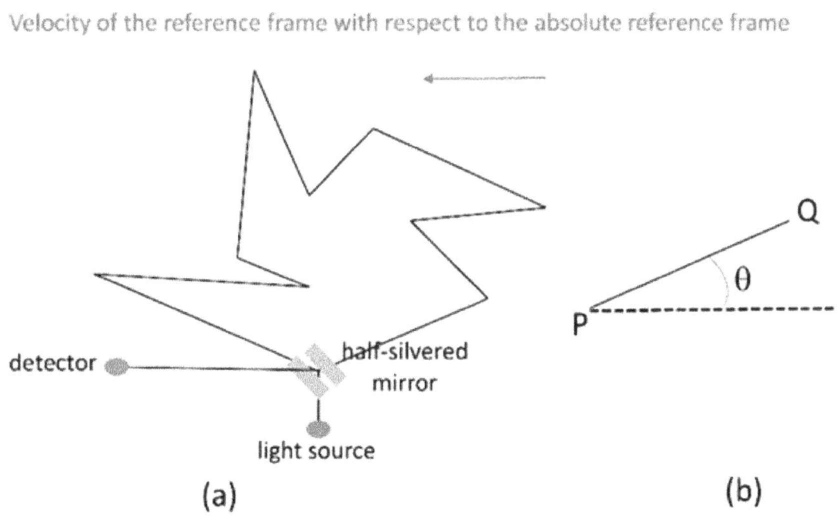

Fig. 17.3. *(a)* A light source generates two coherent beams of light separated by a half-silvered mirror. Therefore, the two rays interfere with each other after having moved in opposite directions through the same path determined by a large number of appropriately oriented mirrors. *(b)* Every PQ segment of the optical path of figure *(a)* forms an angle θ with the velocity of the "absolute" reference system, i.e., with the æther, as indicated by the arrow and by the dotted segment through P. See text.

The result of this experiment must however always be rigorously *zero*, as any segment - which we will call PQ - of the optical path of the two rays "sees" the "æther wind" with its component along the direction shown with hatching in Fig. 17.3. But when you add up all the changes for the "æther wind" that the ray "sees" along all the segments of its optical path, the total sum is zero as the optical path returns *exactly* to the starting point – i.e., to the semi-transparent mirror.

In contrast, a hypothetical, physically significant experiment would use two *independent* light sources, located at different points H and K, respectively, and at a significant relative distance HK. Suppose that such an ideal experiment can actually be carried out, based on the interference between two independent and *strictly coherent* light sources. Send a beam of light from H towards K and study the interference of the light source H with the light source K. If the distance HK is sufficiently large, in principle, the effect should be detected related to the component - compared to the "absolute" reference system of the æther - of the speed of the two reference systems of H and K, respectively.

In reality, the greatest difficulty is just that of having two identical and coherent sources (assuming it is possible) to be used at two points H and K far apart.

In any case, the misunderstanding of the incorrect interpretation of the Michelson and Morley experiment still exists in all physics texts, and it will exist until it is realized that the Lorentz factor simply cannot depend on $\mathbf{v}^2$ and thus be independent of the direction of **v**. See Chapter 20 and Chapter 21.

Georges Sagnac

The French physicist Georges Sagnac (1869-1928) was a well-known scientist, well-connected in the establishment. He never believed in relativity, and was perfectly aware of the meaning of his experiment, which we describe below - even if he had not noticed Einstein's aforementioned "mistake" (see Chapter 16). His close scientific collaborator was his beloved friend Pierre Curie (1859-1906) - who in 1903 was awarded the Nobel Prize in Physics - together with his wife Maria Salomea Skłodowska [better known as Marie Curie or Maria Skłodowska Curie (1867-1934) the first woman to be awarded the Nobel Prize], and Antoine Henri Becquerel (1852-1908) "*in recognition of the extraordinary services they have rendered through their joint research on radioactive phenomena discovered by Professor Antoine Henri Becquerel*".

But on April 19, 1906, Pierre Curie died in an accident in Paris, leaving his wife a widow at just *38* years of age. Pierre Curie, while crossing a busy street in the rain, slipped and fell under a horse-drawn carriage. One of the wheels went over his head, fracturing his skull. He died immediately. In April 1995, Pierre and Marie were buried in the crypt of the Panthéon in Paris. This misfortune proved devastating to Georges Sagnac. The experiment of Georges Sagnac is shown in Fig. 17.4, his original plot.

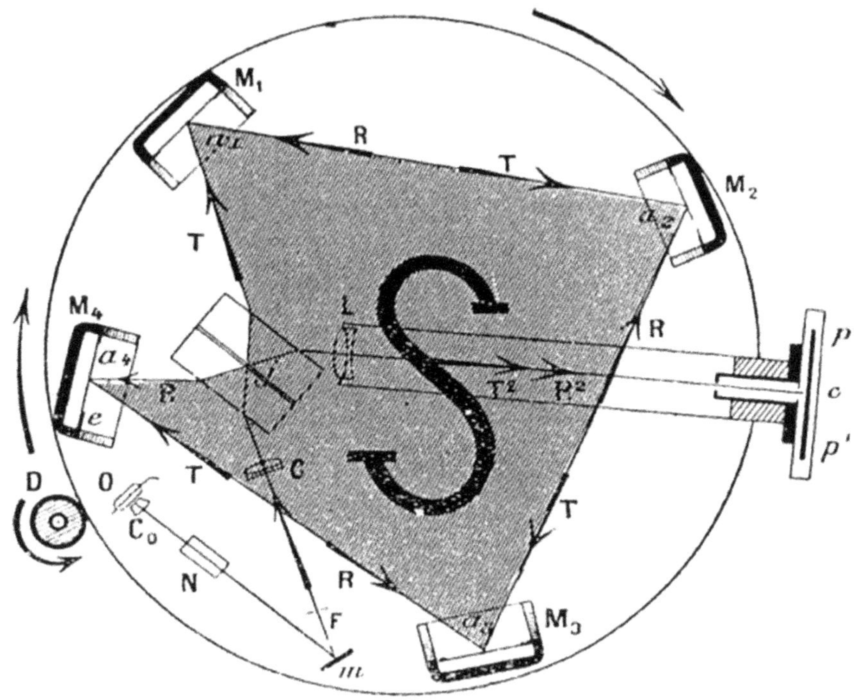

Fig. 17.4. Georges Sagnac's experiment. A light source O sends a light ray through a semi-silvered mirror *j*. The beam is split into two beams that go to mirrors M_1 and M_4 and run in opposite directions along an optical path placed on the rotating platform. Finally, the two beams interfere and are monitored by a photographic plate placed in PP'. See text.

Sagnac's experiment - performed in 1913, i.e., *8* years after Einstein's historic memory - was clearly successful. In fact, in Fig. 17.4 the geometry of the entire trajectory of the two light rays is strictly constant over time.

Therefore, if the platform is not placed in rotation, the result must be strictly zero for the same reasons discussed in Fig. 17.3. On the contrary, if the platform rotates, the orientation of every segment of the path of the light ray varies over time - compared to the unknown "absolute" reference system of æther. So the shortening/expansion of the wavelength is different when it occurs at different instants of time along the same segment of the optical path. The path, by the two rays of every segment (such as PQ in Fig. 17.3) occurs at slightly different times. However, no exact compensation occurs, and the overall effect summed across the entire path *cannot* be zero. Georges Sagnac - in his original experiment - rotated the platform once clockwise and once anticlockwise, and compared the results. Furthermore, in general, the effect should be greater the greater the rotation speed of the platform. It should also be noted that the rotation of the platform occurs with respect to any reference system, without requiring a relativistic speed, and the same description is therefore valid in the framework of both classical and relativistic electrodynamics.

The experiment was a brilliant idea – even though with no acknowledgement by the establishment that preferred to pursue the chimera of the false interpretation of the Michelson and Morley experiment, which had been performed for the first time in 1881, a good *32* years earlier.

Note that three simultaneous Sagnac experiments, conducted at three different locations on the Earth's surface by means of horizontal rotating platforms, can well detect the component of the Earth's rotation axis along three different directions - every direction being perpendicular to the Earth's surface at every venue of the experiment. Furthermore, when a platform can also be operated in a vertical position, three Sagnac experiments can be operated simultaneously at the same site with different platform orientations, and are equally effective. This achievement is of enormous importance.

In other words, one can actually measure the "æther wind".

However, after the passing of Pierre Curie, Sagnac was "alone". The young French theoretical physicist Paul Langevin (1872-1946) – *3* years younger than Sagnac, and who was among the coweorkers of Pierre and Marie Curie - could certainly have helped him. Instead he preferred to propose and devote himself to the explanation of the twin paradox, etc. Furthermore, Georges Sagnac - according to his brother on the occasion of his *post-mortem* celebration - in the last years of his life had lost the luster of his youthful thought, and was unable to defend his interpretation. He was thus progressively marginalized, and his experiment was placed in a limbo of unimportant "curiosity".

But physics, as a "*reading of the book written by Mother Nature*" does not rely on paradigms, on "generally recognized truths", on "now well established truths" etc. The research proceeds regardless of what the establishment deems to be an "established truth". Thus, industrial research proceeded in the development of Sagnac's experiment, using the potential provided by the most recent discovery of lasers and optical fiber... and the most striking results were yet to come.

The laser (acronym for *Light Amplification by Stimulated Emission of Radiation*) is a device that is capable of emitting a monochromatic and coherent beam of light. The discovery, after previous discussions involving other scholars, is traced back to 1950, by Alfred Kastler (1902-1984), a French physicist born in Alsace, when Alsace was part of the German Empire. Kastler proposed the method for optical pumping, which was later confirmed experimentally two years later. Kastler was later awarded the Nobel Prize for physics in 1966.

Optical fiber is now a well-known concept in telephony, or in medical applications, etc. The name "optical fiber" indicates a material made up of glassy or macromolecular (polymeric) filaments that conduct light. The filaments - with a coating diameter approximately the size of a hair and very low weight - are made so that they can conduct light inside them. Their discovery was the result of long studies, and today they find a number of applications that seem endless.

The Fiber Optic Gyroscope

Sagnac's experiment was repeated more recently using a laser source and an optical fiber with several coils (Fig. 17.5). The greater the number of coils in the spiral, the greater the amplification of the effect and the greater the sensitivity of the instrument.

We must now remember a mechanical device, a classic of physics that has been well known for centuries, namely the *gyroscope*. This is a device that maintains a fixed orientation with respect to the system of fixed stars - and the *gyroscope* is essential for, e.g., maintaining the trajectory of a rocket, or of a space probe, or of a satellite, or of a nuclear-powered submarine that remains underwater for years without any astronomical reference, etc. The drawback is the weight of this instrument, which cannot be miniaturized when high precision and sensitivity are required.

The mechanical device - a sort of "spinning top" - is perhaps the most complicated mechanism of the dynamics of solid bodies. It was invented in 1852 by the French physicist Jean Bernard Léon Foucault (1819-1868) as part of his studies on the rotation of the Earth (he is the physicist well known for the "Foucault pendulum"). If a gyroscope is installed on a suitable suspension (gimbal) that allows a rapidly rotating wheel to orient itself freely in the three directions of space, its axis will remain oriented in the same direction even if the support changes orientation. The relevant mathematical interpretation is very intricate and is based on the formulas fully developed by Leonhard Euler (1707-1783), a Swiss mathematician, physicist, and astronomer, born in Basel, considered the most important mathematician of the 18th century, and one of the greatest in history. As already noted, this marvelous mechanical device, however, to be reliable, must necessarily have a considerable weight.

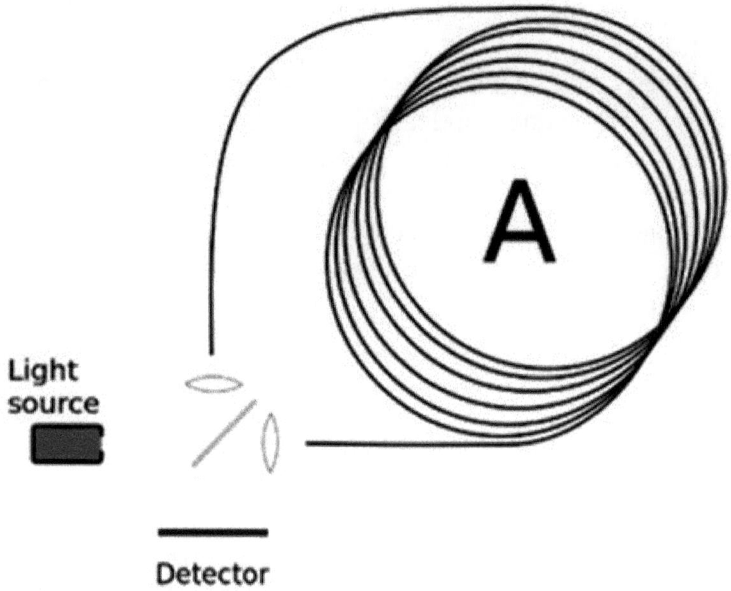

Fig. 17.5. The Sagnac experiment can also be carried out using a laser and a multi-coil optical fiber, to increase the sensitivity of the device. This is the primary mechanism of a fiber optic gyroscope. See text.

Starting in the 1970s, Sagnac's experiment with optical fiber allowed the construction of a very efficient gyroscope with a greatly reduced volume and weight (as in Fig. 17.5). Georges Sagnac has thus taken a great revenge on his detractors!

Indeed, the story of the denigration - a sort of "*damnatio memoriæ*" - of Georges Sagnac is one of the *saddest chapters* in the recent history of physics, and Georges Sagnac has not yet had his due recognition! Let us therefore spend a few words on this truly very sad "episode" which says a lot about the role of the falsehoods propagated by incorrect (and self-interested) communication.

All of this was well highlighted and documented in a 1993 paper by Donald Angus MacKenzie (1950-) who - it should be stressed - is Professor of Sociology at the *University of Edinburgh*. Therefore, MacKenzie was able to observe the detailed historical development, while he was impartial to natural scientists, or to physicists, or to philosophers, or to technologists, or to industrial engineers and/or managers, or to researchers, or to marketing or management specialists, in classified military or mass media research. Instead, MacKenzie, Professor of Sociology, focused on the implications arising from all aspects, and on the reactions and impact on all components of society.

There is no need here to report details that need to be linked to the history of the last years lived by Georges Sagnac. Let's just say that industrial research - often motivated mainly by military-type applications - has moved forward a lot, regardless of the "subtle" disquisitions of theorists and philosophers who are often more concerned with the "consensus" of colleagues and their own "careers" than with research of truth. Rather, all of this brings us to the next point.

Physicists and Philosophers – Debates and Misunderstandings

The long, and still not quelled, debates between physicists and philosophers (in fact, philosophers cannot accept the paradoxes deriving from Einstein's "mistake", etc.) have been very well illustrated in a large volume by Jimena Canales, published in 2015 by *Princeton University Press*. Jimena Canales (1973-) is a Mexican-American historian of science and author with a background in physics and engineering. From 2013-2017 she was Professor of History of Science at the *University of Illinois* at Urbana-Champaign.

This long - and never solved - controversy between physicists and philosophers is symbolically referred to as the controversy between Einstein, the scientist, and the French philosopher Henri-Louis Bergson (1859-1941), even if the debate has involved - and is still involving - many specialists starting from the realm of the Church to the realm of physics and philosophy. The differing opinions have sometimes been even bitterly debated.

This still unresolved controversy concerns the definition of "time". In reality, physicists and philosophers have different objectives. The philosopher - in a certain sense - reminds us of the presumption of human beings, dating back to Aristotle. On the contrary, the role of the physicist is much humbler and "simpler", since he only wants to give a "technical", quantitative and rational interpretation of observations, consistently with the empirical constraint.

Unlike the philosopher, the physicist has no presumption of understanding every "last" detail, or every "fundamental" law, etc. But in this case, evidently, the physicists construct their arguments on Einstein's "mistake", and their position is therefore clearly untenable.

Consider, however, that this decades-old debate or dispute - still not quelled - has allowed many scholars to foster their prestigious careers - as well as to create the "myth" of Einstein's "genius" - while poor Georges Sagnac was unduly left by the wayside. It's a really *very sad* story.

Chapter 18

Quanta and “*Statistical*” Models

“The professional duty of a scientist confronted with a new and exciting theory is to try to prove it wrong. That is the way science works. That is the way science stays honest. Every new theory has to fight for its existence against intense and often bitter criticism. Most new theories turn out to be wrong, and the criticism is absolutely necessary to clear them away and make room for better theories. The rare theory which survives the criticism is strengthened and improved by it, and then becomes gradually incorporated into the growing body of scientific knowledge.”

John Edward Dyson (1941-2010)
[*Infinite in all directions*, 1988, Penguin, London, pp. 258-259]

....

In our logical journey, we have highlighted how Newton’s formulation - later extended to the domain of electromagnetism with Maxwell’s equations - is very effective for engineering applications, whereas Maxwell’s equations have introduced a new unexpected aspect, namely that of the finite speed of the electromagnetic interaction, which was found to be identical to light speed. This generated paradoxical situations which – after a great logical labor – led to the theory of relativity.

We then stressed how a correct formulation that avoids paradoxes must be based on completely different logical premises, and we proposed new axioms, which seem to be the minimum possible required in order to avoid paradoxes. In any case, our approach strictly avoids any continuous function (see Chapter 2) and is based on discrete quantities – i.e., everything in nature is quantized.

The new axioms, however, are *per se* in contradiction with the Newton/Maxwell formulation, which should rather be understood as a regular trend, i.e., the result of a suitable average obtained by interpolating the discrete values with continuous functions. This “statistical” approach has allowed for very notable computational developments, extremely useful in engineering, even though at the expense of unavoidable paradoxes.

But when we entered the domain of the atom monad - the phenomena of which are known to us through spectroscopy, which is based on the photons emitted by an atom monad by the "slingshot effect" - we clearly found that the Newton/Maxwell formulation can no longer hold.

It took a "flash of genius" on the part of Schrödinger, who invented - in a more or less empirical, intuitive, and apparently very strange way - a sort of interpolation on the observations, made up of quantized photons, thanks to a bizarre procedure composed of "complex" functions (i.e., functions having a "real" part and an "imaginary" part; see Chapter 9). As unbelievable as it is, the Schrödinger equation works, and this is indisputable. However, its physical interpretation has never reached unanimity in the scientific community.

Let's say then that it seems very reasonable to guess that the "statistical" procedures - used to interpolate on discrete phenomena that occur within monads of various sizes - are not identical for every monad. In particular, in ordinary life, we know the "statistical" methods that we apply to lotteries, or to the study of "exit-polls" during political or administrative polls, or to all games in general, such as playing cards, etc. But these statistical methods are correct because we know what the laws are that govern the "electoral body monad," or the "lottery numbers monad," or the "playing card monad," etc., which we can well know from the inside. On the contrary, nothing - nothing at all - authorizes us to use the same "statistical" criteria in other monads, especially if observed only from the outside.

In other words, the phenomena we study in laboratories represent a reality that we observe on certain scales and – often – from within monads. In this case, the Newton/Maxwell formulation was built *ad hoc* to make the best use of these phenomena for engineering applications.

However, when we tried to extrapolate the same methods to the atom monad, which is observed only from the outside, we had to realize that the same criteria used for Newton/Maxwell could not be suitable. Schrödinger's formulation had to be used.

Finally, when we tried to apply Schrödinger's formulation to the subatomic world - i.e., to the physics of the atomic nucleus - we found that the Schrödinger's formulation is certainly unsuitable for this purpose. We had to invent a new methodology. The problem already required rivers of ink, and it is still unsolved. This is one of the enigmas that we aim to solve in our journey.

In any case, the panorama that we find is - perhaps - less difficult than it might seem. We try to summarize the state of the art in Table 18.1. Some "technical" terms are explained in more detail in previous or upcoming chapters. That is, Table 18.1 is a sort of "index" of the different evidence that is provided by nature and which is relevant to clarifying many aspects of the formulation discussed here. In other words, Mother Nature is very generous in letting herself be explored and interpreted. It is enough that we have humility - in the strictly Franciscan sense of the term - and not expect phenomena to occur according to our preconceived conceptual schemes.

In summary, the meaning of Table 18.1 will be clarified by gradually exploring the Universe, according to the journey outlined in the different chapters.

Table 18.1 - A synthesis of domains and statistical methods

Name	Domain	Observations	Recalls	Type of statistics requested	Comments
interaction κ	sub-particles	virtually all known experiments performed using high-energy particle accelerators	a huge amount of standard literature	no	The term κ interaction derives from the structure of the photon and charged particles and antiparticles. It is necessary to evaluate how the whole world of speculation on quarks etc. fits into this context, and how the provisional name which is suggested here can be suitably improved
chirality	photons, charged particles, Compton scattering	well known standard experiments	see some references in various chapters	no	The concept κ refers to the structure of the photon, such as a double helix, one helix in the α-world and one in the γ-world (referring to time and anti-time,

					respectively; see Fig. 10.4). Each helix is composed of κ *elements* (electrical or magnetic in nature), and a missing κ *element* determines the structure of a particle charged with both signs, including the spin and the magnetic moment (see Chapter 12 and Chapter 13)
Schrödinger (quantum mechanics)	atoms and molecules	standard experiments related to spectroscopy and chemistry	all standard literature, plus Chapter 7 concerning the golden ratio and fractality inherent in the electron cloud surrounding atomic nuclei	Yes (Schrödinger equation) ed entanglement	See various chapters
Newton & Maxwell (classical physics)	electrical and magnetic interaction between atoms and molecules	crystallography and "cold" matter	standard literature on crystals, including ices	Yes (Newton's principles and Maxwell's laws)	This is the domain of the classical equations of dynamics, which are based on the continuity approximation. See various chapters

Newton & Maxwell (classical physics)	our entire macro world	all experiments related to Newton's principles of dynamics and to Maxwell's law	the entire domain of engineering (non-quantum phenomena)	Yes (Newton's principles and Maxwell's laws)	*Idem*
gravitation	the Solar System and astronomy of the Universe	observations of the dynamics of planetary objects, stars, galaxies, etc.	the entire domain of celestial mechanics (planetary, stellar, galactic, etc.)	(no)	*Idem*. See various chapters
chirality applied to organic chemistry	organic chemistry	all experiments related to organic chemistry	the whole realm of organic chemistry	Yes (antifragility)	See Chapter 26
chirality applied to biological structures	spirals in molecular biophysics	all experiments related to molecular biophysics	the entire domain of molecular biophysics	Yes (antifragility)	See Chapter 26 and Chapter 27
endemic origin of life, evolution, survival, recovery of damage to living structures, generation of new living beings to replace elderly individuals (autopoiesis)	spirals in molecular biophysics	the whole realm of biology	the whole realm of biology	Yes (antifragility)	See Chapter 27 and Chapter 28
solar wind and stellar wind	self-confinement of the solar wind and filamentary structures	interplanetary probes and astrophysical observations	numerous articles of the standard literature	(Yes) *MHD* plus Cowling dynamo	See Chapter 26
star alignments	stellar alignments within	astronomical observations	numerous images reported	No (Cowling dynamo)	See Chapter 26

	galaxies and jets		on the web			
galaxy formation	spiral branches of galaxies, and of imploding matter that generates a planet	astronomical observations	numerous images reported on the web	No (Cowling dynamo plus chirality)	See 26	Chapter
super galactic clusters	alignments of galaxies within super galactic clusters	astronomical observations	numerous images reported on the web	No (Cowling dynamo plus chirality)	See 26	Chapter

Before moving on to the discussion of all these different domains of observed phenomena, it is, however, necessary to illustrate - as synthetically and as clearly as possible - the mathematical aspect of the description of phenomena, according to the scheme proposed here, and alternative to the original Newton/Maxwell formalism.

It will therefore be inevitable to report some formulas here (without exposing the derivation process), explaining the physical meaning of the different terms, and how they differ from the Newton/Maxwell treatment. This will be the only part in which the reader will be asked to pay a little patience and attention to mathematical formulas - otherwise almost completely banned from the discussion provided here and based only on basic epistemological aspects.

The next chapter will then be just a brief summary of what has been discussed in the previous chapters, a concise reminder of some essential points, before introducing the mathematical formulation.

Chapter 19

A Brief Summary – From Newton to Schrödinger – Today's Theoretical Physics Paradoxes – Antimatter

"Whenever there is a breakthrough, a really important new discovery, this means that the experts have been proved wrong, and that the facts, the objective facts, were different from what the experts expected them to be."

Sir Karl Raimund Popper (1902-1994)
[*A world of propensities* (1990, p. 33)]

....

Our journey began before Newton. Newton had set order - and freed himself from the misunderstandings of Aristotelian science. He had introduced - with great practical spirit - the bizarre concept of "force" defined by the lengthening of a thread. Then he introduced mass, conceived as a sort of index of "laziness" (called "inertia") of a body (of any composition). Mass was the coefficient of proportionality, which relates the force applied to the change in motion of the object to which the force is applied.

Finally, Newton had "invented" the law of universal gravitation to explain the impressive empirical laws on the motion of the planets, discovered by Kepler's incredible and extremely accurate analysis. Newton identified, among other things, the *inertial* mass previously defined with the *gravitational* mass that enters the law of universal gravitation. See Section 4.

But there were some logical flaws in Newton's formulation, they were true original sins (see Chapter 5), the most salient of which was, perhaps, the assumption that the gravitational signal propagates instantaneously with infinite speed.

This logical approach had enormous advantages. It triggered the *Enlightenment*, with all its social and historical implications. All engineering successes up to the beginning of the 20th century were determined by this "lightning" synthesis by Newton!

Among other things, in this way, the concepts of the various forms of energy (potential, kinetic, etc.) were introduced - as simple logical derivations from Newton's principles. In any case, the Universe was conceived as a single whole characterized by two large independent and very distinct reservoirs, that of mass and that of energy.

It was natural to apply the same logical schemes to electromagnetism as well, and - by generalization - thus arriving to Maxwell's equations. But a new fact was found. The electromagnetic signal had to propagate at a finite and easily calculable speed. With enormous surprise it was found that this speed coincides exactly with the speed - measured independently - of light! Therefore, light is an electromagnetic wave, defined by an appropriate wavelength – or by an appropriate frequency (which is the same). This finite speed introduced, however, unexpected difficulties and paradoxes. See Chapter 5. So, something needed to change!

At this point Einstein stepped in, also committing a logical mistake (see Chapter 16). As far as our journey is concerned, however, let us remember three fundamental contributions due to Einstein.

As a first contribution, Einstein hypothesized that – in our observed world – nothing can travel faster than the speed of light. It was certainly a very brilliant, "lightening" idea.

As a second contribution, Einstein proposed the formula $E = mc^2$ and it was clear that mass and energy could not be part of separate reservoirs. In fact, mass and energy are considered to be the same component of the Universe, measured with different units of measurement (a bit like measuring the weight of an object in grams or metric tons), despite being physically coincident entities, but manifested differently. Mass and energy are expression of a unique entity, and for convenience we call this entity "emp" (acronym for primordial energy and mass).

In fact, Einstein did not like to specify the genesis - or source - of the ideas he reported in his papers. He took this formula - much to his amazement - from a very similar formula that appeared in a *60*-page memoir published in 1903/1904 in the *Atti del Reale Istituto Veneto di Scienze, Lettere ed Arti* - presented by Olinto De Pretto (also with much amazement) at the *Reale Istituto* on 29th November 1903. Olinto De Pretto was a teacher at the *School of Agriculture* of the *University of Milan*. The memoir was subsequently published in the *Atti* in February 1904 (vol. LXIII, part II) with the title "*Ipotesi dell'etere nella vita dell'Universo*" (Hypothesis of the æther in the life of the Universe). Many circumstances make it more than likely that Einstein read this memoir and published the idea as his own. Olinto De Pretto - *22* years older than Einstein and Director of a family business that still exists - could not assert his priority because he was killed in 1921 by a woman who held him responsible for her husband's economic difficulties. However, the importance of formula $E = mc^2$ was only recognized years later.

At the time De Pretto (like Einstein) was not an important figure in the academic establishment. De Pretto, however, was accepted as a member of the *Accademia dei Lincei*, then chaired by Ernesto Mancini, who was also a member of the *Royal Society of London*. De Pretto's memory was thus favorably received and included in 1906 in the Catalog of international scientific literature of the *Royal Society*. In any case, Einstein had the merit - despite his mistake (see Chapter 16) - of making the formula better known to the international scientific community.

The third salient and rigorous contribution due to Einstein - and used in our discussion - concerns what is here called the "ABC theorem" (see Chapter 3), which was used by Einstein even if he did not give it any name.

Let us now leave aside all the debates - which lack a rigorous scientific basis due to Einstein's mistake - including the discussions on general relativity, which must be considered as deriving from a formal aspect of representation of physical reality, depending on an arbitrary choice of the system coordinates by the observer.

Instead, let us recall the experimental discovery of the existence of antimatter (see Chapter 4). The discovery, which took place in the early 1930s, was immediately evident and indisputable, even if it immediately appeared paradoxical.

To escape this paradox - which still exists - it seems that there is no alternative other than supposing that the "absolute" world is the timeless one. It is thus more than natural to suppose that we must distinguish three different worlds, which we can call the α world, the β world, and the γ world (see Fig. 4.1). The truly "absolute" world is the β world, with no time. Occasionally, a "*materialization*" takes place - and so from the world β arise objects placed in two mirrored worlds, respectively, in the world α characterized by "time", and in the world γ characterized by "antitime", or by a time with the opposite direction and where mass, energy, and emp are negative. The world α and the world γ are mirror images of each other, and mathematically they are expressed - when interpolated with continuous functions - by complex conjugate functions (i.e., they are functions with an identical real part and an imaginary part with opposite signs).

In this respect, it should be remembered that the Schrödinger equation uses a continuous and complex function. However, owing to the empirical constraint, only the amplitude can be observed.

When objects from world α meet with objects from world γ the reverse process occurs, i.e., the objects are "*annihilated*" and return to being part of the world β, with no time.

We can thus assume that there are galaxies of matter that are part of the α world, and antigalaxies of antimatter that are part of the γ world. In the world α the entropy increases, while in the world γ the (negative) antientropy increases (see Chapter 4). When galaxies and antigalaxies annihilate and enter the β world, without time, it happens that entropy and antientropy cancel each other out, and start from zero. Thus, the well-known and disturbing *paradox of entropy* - on which rivers of ink have been spent - is solved.

It then appears very reasonable to suppose that Einstein's hypothesis - that nothing can move faster than the speed of light - can be simply generalized as follows.

A photon observed in the β world can have no velocity - because "time" or "antitime" is not defined, and it is not possible to define any velocity. Furthermore, the photon is made up of two "spirals" (Fig. 10.4), one wrapped in the α world and the other in the γ world. Therefore, the photon intrinsically has no component with a prevalent time or antitime direction. Furthermore, the photon is a volley of miniphotons, and every miniphoton has an intrinsic content, which is well defined and measurable, of emp and antiemp, depending on whether the photon is considered part of the α world or the γ world.

In contrast, a photon observed in the α world where a velocity can be defined, will always be seen to move with the speed of light. Similarly, a photon observed in the γ world, where it is possible to define a velocity referring to antitime, will always be seen to move with the speed of light, even if referring to antitime.

Concerning every particle (in the α world) and every antiparticle (in the γ world), they are "*degraded*" forms of emp because they cannot move at the speed of light – while the speed of light is an exclusive property of photons, i.e., of miniphoton volleys that are not "*lame*".

The interactions among particles, antiparticles, photons, miniphotons, etc. must be interpreted by means of Feynman graphs, appealing to the interactions of "*elements* k_E" and "*elements* k_H", and considering volleys both complete and "*lame*", etc., as detailed in Chapter 13, Chapter 14, and Chapter 15.

However, we ask ourselves whether and how we can generalize the Newtonian concepts of conservation of mass and energy, and of conservation of "momentum", and of "angular momentum" (see the description of interaction between billiard balls of Chapter 4 and Chapter 12).

Note, however, that these "conservation laws" are in themselves a complement, an approximate appendix, of Feynman's graph-based formulation - and therefore represent a way of making a "global balance" of a phenomenon that should be described in a rigorous and more detailed manner, only by considering one "*element*" at a time.

In this respect, consider that the conservation of energy is easily translatable into conservation of the total quantity of emp - given that emp is a well-defined and quantized component of the physical system.

Conversely, every quantity that derives from the velocity of a particle or antiparticle refers to a continuous parameter (the velocity) which has a descriptive value within a reference system of space and time (or antitime). However, everything (such as a speed and its derivatives) that is not quantized but continuous is *per se* approximate, averaged, "statistical", and therefore it has a less reliable value in an axiomatic framework.

Once these prerequisites have been clarified, the synthetic - and broader perspective - formulation is presented in the following chapter. We will see that the conservation of energy can easily be generalized to the conservation of the amount of emp before and after the interaction between two objects.

In contrast, the concept of momentum - which in Newtonian physics is the product of mass and velocity (see Chapter 4) - cannot be easily generalized, since this would violate the ABC theorem. This is important because it will be seen that the similar formulas proposed by Einstein violate the ABC theorem and must certainly be rejected.

In any case, we will only show the final formulas, excluding the related derivations, which require only the formal use of the language of mathematics, and which for clarity we have excluded from our journey. In any case, our journey becomes increasingly fascinating! We warn the reader that the next two chapters are – in a certain sense – the heaviest obstacle in our journey. The subsequent chapters will be much easier to read.

Chapter 20

Contraction and Expansion of Units

"Here's to the crazy ones, the misfits, the rebels, the troublemakers, the round pegs in the square holes … the ones who see things differently – they're not fond of rules … You can quote them, disagree with them, glorify or vilify them, but the only thing you can't do is ignore them because they change things … they push the human race forward, and while some may see them as the crazy ones, we see genius, because the ones who are crazy enough to think that they can change the world, are the ones who do ... Stay hungry! Stay foolish!"

Steve Jobs (1955-2011)
[Commencement speech to the graduates of *Stanford University* in 2005]

………………………………………

This chapter summarizes the concepts developed in what precedes, and - beyond the conventional mathematical formalism which is included here for the strictly necessary minimum - this chapter perhaps requires greater commitment from the reader, who will have to follow the entire argument very carefully. Let's proceed by ordered sequential steps.

The ABC theorem, the definition of the measurement units for space and time coordinates – and the contraction/expansion of time and length

In apparent disagreement with Einstein's classical treatment (see Chapter 16), we recall that an absolute reference system or æther certainly exists (see Chapter 3). Furthermore, let us remember that an observer may also be incapable of recognizing the absolute reference system. The important thing is that – whether recognized or not – the absolute reference system (i.e., æther) exists.

3D space is a sort of "shelf" where we insert and describe the phenomena we observe. A space is defined by coordinated axes (whether linear or curved, arbitrarily chosen, as we like). Suppose we want to use orthogonal linear coordinate axes, i.e., a reference that is conventionally called Euclidean (since it is based on the well-known elementary geometry of Euclid). We must therefore first define what we can physically assimilate to a straight line. In fact, the abstract concept of a straight line that we have in mind is one thing, but what a straight line is in nature is another.

What in nature is there that represents a straight line better than a ray of light? Realistically nothing, nothing at all! Certainly a ray of light is deflected in the vicinity of an object of great mass, i.e., of great emp content. In this respect, we recall the astronomical experiment mentioned in Chapter 15, which showed how the gravitation of the Sun deflects the light of the stars observed during a total solar eclipse. We can say that a ray of light travels in a straight line where gravitation is not a significant source of disturbance.

On the other hand, the entire Universe is home to "*messengers*" who generate a gravitational interaction, and the above experiment shows that, in fact, light is affected, as one should expect, by gravitational interaction (see Chapter 15).

Furthermore, it just doesn't seem possible to hypothesize something alternative to a ray of light to define a "straight line". Indeed, the concept of "straight line" is a mere abstraction of our mind. We can consider a ray of light to be a "straight line" only because we have no alternative at all. See Figs 3.5 and 3.6. Due to the empirical constraint, we are not very different from the ants in Fig. 3.6.

In conclusion, only a ray of light is the only instrument available suitable for defining a "line" in natural reality that we call a "straight line". 3D space is therefore not only "absolute", but is also isotropic, i.e., a ray of light propagates by our assumption in a straight line in all directions - and this is true for the sole fact that we possess no other observational instrument for defining a straight line to compare with a light ray.

Einstein had tried to circumvent this obstacle - due to the empirical constraint - by imagining a rigid "ruler" (see Chapter 16). But no rigorous argument can rely on any "ruler", because "solid" objects cannot be included in any axiomatic approach. In fact, the vast fraction of the Universe is excessively hot to contain matter in a "solid" or "rigid" state. Therefore, a rigid "ruler" is a pure conceptual abstraction, no more and no less than an abstract straight line, which cannot have an objective counterpart in reality.

In fact, the presumed constant speed of light c is the only tool suitable for defining a unit of time. To this end, e.g., consider the frequency of a chosen line of the spectrum of a given element, or consider the duration of a complete oscillation of a given electromagnetic wave that has been conventionally chosen as the unit of measurement of "time", and agreed upon by all observers, either moving or not.

In order to define the elementary unit of length in space, it is then sufficient to multiply the time unit by c - as a velocity is a length divided by a time, and therefore a length is a time multiplied by a velocity.

That is, "time" is just a "technical" definition identified by a phenomenon that displays a regular cycle. That is, we identify "time" with a "clock" of formal reference - being a "technical" instrument arbitrarily chosen by every observer. That is, the "clock" is just a simple tool used by every observer to compare different observations, without posing philosophical implications of any kind. The important thing is rather that the observer - like any living being - has the perception of the flow of "time" precisely because he has the possibility of observing a cyclical phenomenon - such as, e.g., an electromagnetic wave with an assigned frequency (see Chapter 9).

Finally, it is important to emphasize that every observer always operates as an *independent* subject from other observers, through his personal reference system (principle of relativity) and through his personal "clock". Only *after* completing his experiments and observations, every *independent* observer can communicate with other independent observers, who can eventually repeat the same experiments. In this way, different observers can compare their interpretations of natural phenomena.

That is, contrary to what was assumed in Einstein's classical argument, *there is no need* for different observers to *synchronize* their respective "clocks" (see Chapter 16). Rather, we consider phenomena only within the "absolute" reference system, while every independent observer observes and defines the "*simultaneity*" of two observed events only by means of his personal operating horizon and of his "clock", consistently with the principle of relativity.

"Time" is not absolute. But "time" is one and only one in world α. The "time" is only a "technical" detail referring to a reference "clock", which is autonomously defined by every *independent* observer in his personal reference system in the α world. Communications between different observers in world α concern only their respective conclusions regarding the laws of nature – or regarding the comparison of the respective results of experiments which can reasonably be assumed to be repeated under apparently identical conditions.

That is, it is necessary to distinguish two different aspects: *(i)* the definition of the operational framework, the reference system, and the units that are autonomously defined and used by every *independent* observer, and *(ii)* what and how the different observers can communicate with one another concerning their respective conclusions and inferred natural laws. For instance, several observers can choose by mutual agreement a certain spectral line of a certain element (e.g., hydrogen) as measurement unit of “time” and length.

Now let’s recall the ABC theorem (see Chapter 3). That is, A, B and C are three independent observers (Alice, Bob, and Charlie, respectively). Let’s consider a Euclidean reference system associated with every observer (see Fig. 3.7, which we re-propose here as Fig. 20.1). Observer B moves relative to A with velocity $\mathbf{v}$ along the $\boldsymbol{x}$ axis of the reference system of A. Observer C moves relative to B with velocity $-\mathbf{u} = -\mathbf{v}$ along the $\boldsymbol{x}$ axis of the reference system B. Therefore, observer C moves relative to A with zero velocity. That is, A and C are stationary with respect to each other and use identical reference systems - even if the origins of their respective Euclidean coordinate systems can be distinguished. Owing to “*realism*” (see Chapter 3) the physical laws inferred by A must therefore be identical to the laws inferred by C.

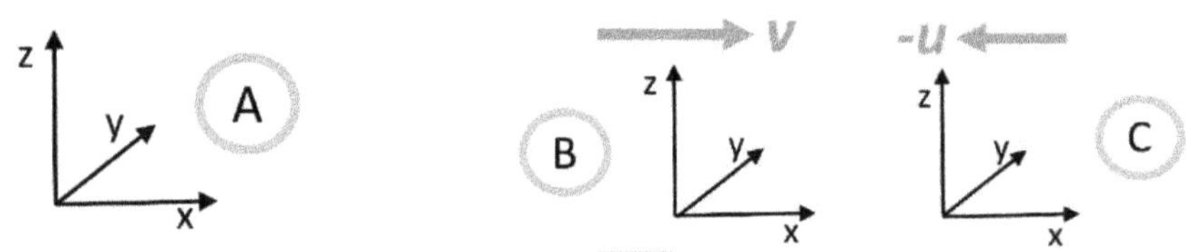

Fig. 20.1. Case of three observers, everyone symbolically represented by his own reference system which - for simplicity - is always assumed to be 3D orthogonal Cartesian. Bob (B) moves with respect to Alice (A) with velocity $\mathbf{v}$. Charlie (C) moves relative to Bob (B) with velocity $-\boldsymbol{u}$, along the same direction as $\mathbf{v}$ but in the opposite direction, as denoted by the minus sign. Therefore, Charlie (C) moves relative to Alice (A) with velocity $\mathbf{v} - \boldsymbol{u}$. Therefore, when $\mathbf{v} = \boldsymbol{u}$, and $\mathbf{v} - \boldsymbol{u} = 0$, Alice (A) and Charlie (C) are stationary relative to each other. So the same phenomenon is observed differently by Alice (A) and by Bob (B). Conversely, Alice (A) and Charlie (C) must observe the same phenomenon identically because, in fact, they are the same observer, possibly placed in different sites. The laws of physics that Alice and Charlie deduce from observations must be identical. If a proposed law does not meet this condition, the law is not correct. This “theorem” was applied by Einstein, even though he didn’t give it any name. For brevity, this argument is called the “*ABC theorem*”.

Suppose that all observers agree on a conventional common choice. For instance, every observer defines his own time and space unit by means of the same line of the hydrogen spectrum, conventionally agreed upon by all observers. That is, observer A refers to his sample of hydrogen, while B refers to another sample of hydrogen in his personal Euclidean reference system, and C refers to another personal sample of hydrogen in his personal Euclidean reference system etc.

Every respective definition of unit of time or space - referred to absolute reality and defined by every observer - can in principle be comparatively more or less contracted or expanded, since no reason requires that no relative contraction or expansion of a measurement unit takes place resulting from the motion of an observer with respect to æther. A contraction or expansion perhaps occurs - or perhaps it does not - and this may be due to some property of hydrogen or some other unknown reason. In any case, "*realism*" (see Chapter 3) only requires that - unlike measurement units - any given phenomenon must always be rigorously univocal, defined in itself, independent of the observer who observes it.

In particular, for a requirement of absolute reality (i.e., of "*realism*", which is the same), it seems reasonable to assume that two phenomena that are simultaneous for one observer must be simultaneous when the phenomena are observed by every other observer (see Chapter 16). In fact, "time" is identical for all observers who are in the world α and no known physical reason requires otherwise. Note that this is true for every pair of observers, regardless of their respective position, or relative motion, or environment, because *simultaneity* (in world α or world γ) is related to "time" or "antitime" *per se*, and "time" and "antitime" are defined independently of the observer, and are *physically unique* in the α world and the γ world, respectively. Every observer simply measures "time" or "antitime" through his personal "clock".

For simplicity, we can assume, e.g., that the two observers are located at a non-excessively large distance from each other, in terms of light years etc. For instance, let's say the two observers are on the same planet. But this is not strictly indispensable. If the two observers are on opposite sides of the same planet, the reference system moves in opposite directions due to the planet rotation. Suppose a disagreement occurs between the two observers regarding "simultaneity". It must be concluded that the personal "clocks" chosen by the different observers, respectively, are physically incompatible with each other. That is, the two arbitrarily chosen "clocks" are based on different phenomena – and it was incorrectly assumed that the phenomenon chosen by one or by both observers was representative of the evolution of "time" within the absolute reference system.

This assumption – of "*simultaneity*" agreement between two observers as shown below – is the subject of crucial disagreement with Einstein's formulation, with enormous consequences.

In any case, the difference must be considered between:

(i) observations carried out, e.g., by Bob in the reference system that Bob autonomously defines using *his hydrogen sample*, and

(ii) observations carried out by Alice when referring to *Bob's sample of hydrogen* and relative to the reference frame with origin in Bob that is defined and considered by Alice - and which moves at Bob's speed relative to Alice's.

Bob's two reference systems – one defined by Bob, the other defined by Alice – that are thus considered, are strictly identical. One frame of reference is used by Bob, and the other frame of reference of Bob is defined and used by Alice, while Alice observes phenomena as Alice can presume that they are observed by Bob by means of Bob's hydrogen sample.

Both reference systems have the same origin and the same velocity relative to the absolute reference system (i.e., æther). However, Alice and Bob may have a different protocol regarding the way they carry out measurements. That is, Alice and Bob need not get the same result. They only have to respect the principle of "*realism*" and independently deduce the laws of nature, according to their respective analysis, by means of a chosen sample of hydrogen.

In other words, Alice performs experiments and uses Alice's *sample of hydrogen*. Likewise, Bob performs experiments and refers to Bob's *sample of hydrogen*. However, for purposes of comparison, Alice also considers how the same events that Alice observed with *her hydrogen sample* can be interpreted whenever Alice refers to *Bob's hydrogen sample*. Therefore, Alice can compare the results that she can infer if she uses either *Alice's* or *Bob's hydrogen sample*.

Consider Alice, Bob, and Charlie (A, B and C). Suppose that – with respect to the absolute reference system "æther" – B moves with respect to A at a velocity $\mathbf{v}$, and that C moves with respect to B at the inverse velocity $-\mathbf{u} = -\mathbf{v}$, as in Fig. 20.1. Alice and Charlie are therefore at rest relative to each other (although the origin of their respective reference systems can be moved a distance along the $\boldsymbol{x}$ axis and independent of time). So, since A and C are at rest with respect to each other, if A and C use the same line of the hydrogen spectrum, the units – which depend only on the frequency of the same line – *must be identical*, regardless of whether that the units are deduced by A using either the hydrogen sample of A or that of C.

Note that two apparently identical hydrogen samples, each operated in a different environment, can possibly respond differently to some unknown local perturbation. However, A only needs to identify a certain line of the hydrogen spectrum – and that line is used to define the units by means of both samples. The rest of the units are totally unknown. Thus, when A compares the two units that she deduces from both A's and C's samples of hydrogen, she must find (as Einstein pointed out) that the two units are identical.

That is, A is an *independent* observer performing an experiment in the reference frame of A. She uses a given spectral line of a given sample of hydrogen of A to define her units. Then, A considers how she can perform the same experiment when B's hydrogen sample is used by A. Therefore, A considers how the units in B's reference horizon contract or expand relative to her units when, instead of the sample of A's hydrogen, A uses B's hydrogen sample.

Similarly, A considers how to perform the same experiment when she uses C's hydrogen sample. Therefore, A considers how the units may be contracted or expanded when switching from B's hydrogen sample to C's hydrogen sample. In any case, the final units, defined from C's sample of hydrogen, *must be identical* to the first units calculated by A from her personal sample of hydrogen.

This result is inherent in the ABC theorem. Einstein used this theorem, even though he did not give it any specific name.

For the time being, we summarize and anticipate in Table 20.1 (where "*Con*" denotes contraction, and "*Exp*" denotes expansion, and ≡ denotes neither contraction nor expansion) the state of the art which is later discussed in detail.

Table 20.1 – Relative units of measurement of the ideal experiment carried out by A, B, and C

	Alice	Bob	Charlie
Alice	≡	*Con*	≡
Bob	*Con*	≡	*Exp*
Charlie	≡	*Exp*	≡

Alice, Bob and Charlie compare their respective units (every observer in a row of Table 20.1 compares the units with the observer in a column, and vice versa). Concerning symbols, "≡" denotes no contraction and no expansion (i.e., $\chi[(\mathrm{v}/c)] \equiv 1]$, using a symbol defined in Chapter 21), "*Con*" denotes contraction (i.e., $\chi[(\mathrm{v}/c)] < 1$), and "*Exp*" denotes expansion (i.e., $\chi[(\mathrm{v}/c)] > 1$). Alice concludes that, compared to her units, Bob uses contracted units. Instead, Bob concludes that, compared to his units, Charlie uses expanded units. Alice and Charlie have the same units. Some details will be seen later.

The final and subtle related key-point concerns the axiomatic definition of "*simultaneity*", as is more fully stressed below.

We do not insist on further details relating to cases in which there are numerous observers who move with respective speed, more or less in a zigzag path. These formal complications add nothing to the final result, which is clear: if two observers are at rest relative to each other, they must find identical measurement units. This is the ABC theorem.

It is important, however, to emphasize a fundamental disagreement with Einstein. Indeed, according to Einstein's classical formulation, Einstein shows that the length contraction occurs *only* along the direction of the relative displacement of A and B (see Chapter 16). Conversely, according to the axiomatics described here, it is clearly shown that the *identical* coefficient of contraction or expansion of measurement units, i.e., the aforementioned χ ratio, must be applied identically to the time unit, and to all 3 length units of the 3 axes of 3D Euclidean space. In other words, it is shown that χ must be the same for all *4* axes of 4D spacetime. *This inference derives in fact from the isotropy of spacetime* which - as we stressed above - is a mere assumption due to our inability to recognize any deviation from isotropy, like the ants in Fig. 3.6.

In summary, in our formulation we formally show that energy, mass, emp and frequency of a photon transform their respective units according to χ, while, on the contrary, both length and time units transform according to $1/\chi$.

In addition, according to Einstein's special relativity, only unit contraction is considered, and never an expansion. This follows (see below) from a dependence on $|\mathbf{v}|^2$. Therefore, by changing the sign to $\mathbf{v}$, the sign of contraction or expansion does not change. Furthermore, as already stressed - according to Einstein's special relativity - the contraction operates on the axis of "time" and operates *only* on the axis of space, which is along the velocity vector $\mathbf{v}$ which represents the movement of Bob with respect to Alice. But, according to Einstein, *no contraction occurs along either one of the remaining perpendicular space axes.*

In fact, the logic of axiomatics, which we consider in our journey, implies that everything – i.e., the length or width of every object, as well as every interval of time - changes overall by the same factor (either by contraction or by expansion), whereas c must remain constant. That is, the geometric shape of every object is equally scaled (contracted or expanded) in every direction, just as it happens in elementary 3D geometry for the relationship between so-called "similar" figures. Furthermore, all time intervals are scaled accordingly by the same contraction or expansion factor. All this is part of "*realism*", as "time" is absolute in the world α and the same must be said of antitime in the world γ.

Conversely, according to Einstein's special relativity, two spatial axes undergo no contraction, unlike the third coordinate axis. Therefore, some relevant physical implications need to be considered. In fact, in Einstein's formulation - which has the logical flaw of the mistake explained in Chapter 16 - the principle of "*realism*" seems to be violated, and the entire topic therefore needs to be suitably reconsidered. That is, the argument that led Einstein to such a surprising property - that only the length unit along one direction is affected by contraction, and not the length units along the other two directions - violates "*realism*". In addition, "*realism*" is further violated by the fact that only contraction and never expansion can occur. This also violates the ABC theorem. On the other hand, Einstein appeals to the ABC theorem, which, however, he later seems to ignore.

In summary, all paradoxical conclusions of the so-called special relativity, such as, e.g., the famous twin paradox (see Chapter 16), seem to arise from an implicit – and undeclared – assumption of violation of "*realism*", which is latent somewhere within Einstein's original argument. In any case, we must not forget Einstein's fundamental mistake (see Chapter 16) and the mistake in the interpretation of the Michelson-Morley experiment (see Chapter 17).

It follows that the Michelson-Morley experiment proves nothing, because one must always expect the null result. The erroneous conclusion - originally deduced when the experiment was performed - arose from the fact that Einstein's erroneous contraction coefficient $1/\sqrt{1-\beta^2}$ depends on an even power, rather than on an odd power of $\beta = |\mathbf{v}|/c$. From this fact it follows that the shortening of the time unit is erroneously considered the same both when a photon travels along one arm of the Michelson interferometer in one direction, and when it travels along the same arm in the opposite direction.

Moreover, in the case of twins, time contracts as the twin on the spaceship moves away from Earth, but time expands equally as the spaceship approaches Earth.

In this respect, let us remember Georges Sagnac's experiment which led to the development of the fiber optic gyroscope described in Chapter 17. Einstein's mistake was found with Sagnac, the correct observational and indisputable response to the Michelson-Morley experiment.

This whole topic is very simple - and we hope that it has been illustrated clearly here. The reader interested in details should refer to the original scientific literature, which is much more extensive than the summary discussion presented here. There are no particular conceptual problems. It's just a matter of considering everything as a "game" of patience and considering every logical step without rushing.

We have thus clarified the problem of the definition of the measurement units of time and space by every observer, and of how these units can undergo variations (expansion or contraction) depending on the motion of the observer with respect to æther. All of this also holds symmetrically in the world γ referring to antitime.

Chapter 21

Conservation Laws
A "*Statistical*" Model and the Principles of Dynamics
A little (though not too much) Mathematics

[I recommend to young researchers] "*never to stop to question, mainly when solutions appear more difficult, and I recommend to learn to disobey if necessary. Innovation is founded on the ability to transgress; obedience itself is not a value … I never liked dogmatism. I broke the sacred texts of oncology pundits, and I searched for new solutions for breast cancer. Many women had suffered unnecessary mutilation if I respected the hierarchy of knowledge. The value of women and men, even other than scientists, also relies on their ability to look up from the slides and watch a little further, taking the side of a better future according to what they can imagine.*"

Umberto Veronesi (1925-2016)
[famous Italian oncologist, known for his contributions on prevention and treatment of breast cancer;
interview published by *Il Corriere della Sera*, April 24th, 2016]

…………………………………………

Definition of "dynamics"

Let's try, in some way, to assess whether the equations of dynamics that are the backbone of Newton's formulation can be connected to the axiomatic and rigorous formulation according to Feynman graphs - which is the target of our journey.

First let's consider *kinematics* - which is the description of the movement of objects. Remember that we must always refer to the absolute system, i.e., æther. In any case, when we consider a macroscopic body – e.g., a billiard ball, a boulder, or a planet, etc. - we must be able to consider it as a "point" endowed with some intrinsic property (e.g., we must be able to consider whether a billiard ball is spinning or not, and similarly a celestial body, etc.). We will then call a macroscopic body simply with the term "*granule*" to distinguish it from the concepts of subatomic physics, which are only typical of the - "*lame*" or not - spiral volleys of our logical journey, which are referred to the micro-cosmos.

Nothing prevents us from defining the distribution - in our 3D Euclidean space - of the speed **v** of every "granule", and this is called "*kinematics*".

We can also consider the center of gravity (or *barycenter*) of a set of "granules" when we assume them to be in approximately a fixed position with respect to one another - as it happens, e.g., in a "rigid" object, such as a building, a meteorite, a planet, etc. The center of gravity is the location of the ideal point that needs to be considered when all the "granules" that compose the macro-object are supposed to be concentrated in a single point. In this case, the resulting macro-"granule" will have a mass equal to the sum of the masses of all the component "granules". For this purpose, the center of gravity is defined as the average of the coordinates of all the component "granules", where every component "granule" has a statistical weight proportional to its emp content (i.e., a component "granule" that has, e.g., an emp content double of another component "granule" has a double weight in the calculation of the average; this is also named weighted average). All of this can be repeated with no problem in our formulation. But these are only mathematical technicalities that are conceptually simple and formally a bit elaborate, yet relatively elementary, and which are reported in every school textbook.

Similar considerations apply when we define suitable mathematical technicalities that permit us to deal with the intrinsic spin of every single "granule", and how this can be inserted into a dynamic equation. But all of this adds nothing to the epistemological discussion of our journey. We will therefore omit all these details that can be found in classical physics texts used, e.g., in secondary schools.

It is, instead, very important to discuss the definition of what, in Newton's formulation, is called "*momentum*". For Newton, "*momentum*" is defined by $\boldsymbol{p} = m\mathbf{v}$, and it is shown that in a collision or scattering between "granules" (such as between billiard balls) both total energy and "*momentum*" must be conserved. That is, the total values added on the two "granules" of both kinetic energy and "*momentum*" must be the same before and after the collision.

According to our formulation, the conservation of energy is immediately reflected in the conservation of the quantity of emp.

As far as "*momentum*" is concerned, let's try to define the quantity

$$\mathbf{p} = \frac{E\,\mathbf{v}}{c^2} = \left(\frac{1}{c^2}\right)E\mathbf{v} \tag{21.1}$$

which coincides with $\boldsymbol{p} = m\mathbf{v}$, where, consistently with the formula $E = mc^2$, instead of the mass m we put the ratio between the emp content E and the square of light speed c^2. That is, $\boldsymbol{p}$ is (apart from the coefficient $1/c^2$) equal to the "flux" of E per unit time.

The concept of "flux" is basic. Consider, e.g., the flow of a river. The flux of water is equal to how much water flows in a unit of time (e.g., in a second, or in a minute, or in an hour, etc.). Thus (21.1) expresses the flux of emp E, per chosen unit of time, which occurs at every given point in 3D space.

To move from the description of phenomena according to "*lame*" volleys or not, to a "statistical" description according to "conservation laws" which somehow recall the treatment of the dynamics of bodies according to Newton, it is necessary to do some mental "gymnastics" more or less daring, which we will try to explain as best as possible. The inspiring principle will, in any case, be that of the maximum possible "simplicity".

Definition of $\xi(|\boldsymbol{v}|)$, of total energy, of kinetic energy, of inertial mass, and of moment of inertia

Let's introduce a function $\xi(|\mathbf{v}|)$ to be defined in greater detail below. Some properties of this $\xi(|\mathbf{v}|)$ can be derived in the following based on a physical discussion. Therefore, the definition $\xi(|\mathbf{v}|)$ is left loose to allow some additional margin in order to account for observations.

In the standard theory of special relativity $\xi(|\mathbf{v}|)$ is the ratio between the inertial mass and the rest mass of a particle, i.e., $\xi(|\mathbf{v}|)$ is the amplification factor - due to the motion state - of the rest emp of one system. That is, this ratio is analogue, referred to the mass, of the symbol $\chi[(\mathrm{v}/c)]$ which is connected to the contraction or expansion of the measurement units of length or of time or antitime.

It is important to emphasize that we assume for now that the relationship between inertial mass and rest mass is characterized by a *multiplicative* factor, i.e., we assume that the inertial mass is obtained by *multiplying* the rest mass by an appropriate coefficient.

On the other hand, we can also hypothesize, alternatively, that the change occurs through an *additive* criterion – i.e., just as if kinetic energy were added to the rest mass.

We will consider both options. For the moment, however, we will refer first to the *multiplicative* criterion, while at a later stage we will consider the *additive* criterion.

Above, while dealing with Table 20.1, we considered the symbol $\chi[(\mathrm{v}/c)]$ connected to the contraction or expansion of measurement units both of lengths and of time or antitime. That is, the symbol $\chi[(\mathrm{v}/c)]$ in Table 20.1 denotes the inverse of the ratio of the length unit $\lambda^{(A)} = \lambda_A^{(A)}$ defined and used by A within of her reference system A, and the corresponding length unit $\lambda_B^{(A)}$, observed and measured by A, using B's hydrogen sample which is used by B to define his units in B's reference system.

However, we remember (see Chapter 11) that the energy of a photon of frequency ν is $h\nu$. That is, $\chi[(\text{v}/c)]$ is the ratio between the energy $[h\,\nu_A^{(A)}]$ of a photon of frequency $\nu^{(A)}$ of the sample A in A units, and the energy $[h\,\nu_B^{(A)}]$ of the photon released by an identical hydrogen sample of B and observed by A.

Therefore, the symbols $\xi(|\mathbf{v}|)$ and $\chi[(\text{v}/c)]$ denote similar quantities (or perhaps are a single ratio). The difference is that $\xi(|\mathbf{v}|)$ concerns *every form* of emp, while $\chi[(\text{v}/c)]$ *only* concerns *the energy of a photon.*

For the sake of clarity, we will use different symbols for $\chi[(\text{v}/c)]$ and $\xi(|\mathbf{v}|)$, although the conclusions we find for $\chi[(\text{v}/c)]$ or $\xi(|\mathbf{v}|)$ are strictly consistent among them. However, due to Ockham's razor (see Chapter 14), since no physical difference between the two types of emp is to be expected, we will conclude that the coefficients $\xi(|\mathbf{v}|)$ and $\chi[(\text{v}/c)]$ should be same. For now, we use different symbols for $\chi[(\text{v}/c)]$ and $\xi(|\mathbf{v}|)$ and we introduce the required formal definitions.

As a further preliminary warning, we point out that, as mentioned above, separating the translational speed **v** and the rotation speed is only an aspect which is of no new physical content, but rather of computational convention with reference to composite "granules". Thus, the discussion of $\xi(|\mathbf{v}|)$ is conducted here by considering only a translational speed **v**, while the discussion of rotational **v**, i.e. of spin, is a subsequent physically non-essential formal complication - which is necessary whenever we lose the details of the processes that occur inside a granule monad.

However, as already stressed, this separation is only a matter of conceptual simplicity, useful for applications, while - when dealing with actual physical reality - it is necessary to reformulate everything, avoiding the reference to macro-objects such as "granules" or "corpuscles".

The formal definitions are as follows. The reader will forgive these details, even if he is not familiar with the symbols of dynamics according to Newton - such as, e.g., the spin velocity of a "corpuscle" denoted by $\vec{\Omega}$, or the moment of inertia, etc. For simplicity, we omit these details, which are standard in high school physics textbooks, and which are not essential to our discussion.

total energy of a granule	$W = E + K_{tr} + K_{rot}$	(21.2)
granule rest energy	E	(21.3)
translational kinetic energy (due to **p**)	$K_{tr} = W - K_{rot} - E = E[\xi_{tr}(\lvert\mathbf{v}\rvert) - 1]$	(21.4)
rotational kinetic energy (due to $\vec{\Omega}$)	K_{rot}	(21.5)
$\chi_{tr}(\lvert\mathbf{v}\rvert) = [W - K_{rot}]/E$	$E + K_{tr} = W - K_{rot} = \xi_{tr}(\lvert\mathbf{v}\rvert)\, E$	(21.6)
$\chi_{rot}(\lvert\boldsymbol{\omega}\rvert) = K_{rot}/E$	$K_{tr} = W - E - K_{rot} = \xi_{rot}(\lvert\boldsymbol{\omega}\rvert)\, E$	(21.7)
rest mass	$m_0 = m/[\xi_{tr}(\lvert\mathbf{v}\rvert) + \xi_{rot}(\lvert\boldsymbol{\omega}\rvert)] = E/c^2$	(21.8)
inertial mass	$m = [E + K_{tr}]/c^2 = [W - K_{rot}]/c^2 = \xi_{tr}($	(21.9)
moment of inertia (instant value)	$I = \lvert\mathbf{\Omega}_{rot}\rvert/\lvert\,\boldsymbol{\omega}\rvert$	(21.10)

Only a few of these quantities are discussed below. It will be sufficient to consider the nature of the different quantities, i.e., *(i)* the frequency υ of a photon, *(ii)* the energy and *(iii)* the mass, or *(iv)* the emp E, of either a photon or a massive (or antimassive) particle. In physical terms, all these quantities are the same physical quantity, even if measured by different units, while all can also be measured by the unit of E.

Therefore, given the above, it follows that

$$\chi(\beta) \equiv \xi(\beta) \qquad\qquad \beta = \lvert\mathbf{v}\rvert/c \qquad (21.11)$$

A key distinction, however, is required when we deal with "*undegraded*" or "*degraded*" emp. In fact, the "*degraded*" state implies the consideration of the rest mass of a particle or antiparticle – while, on the contrary, any reference to the rest mass of a photon implies a conceptual difficulty, as in the β world no speed is defined, therefore it is not possible to define a "rest" state for the photon. On the contrary, in the α or γ world a photon can only move at light speed, and therefore one cannot consider a photon at rest (see Chapter 6 and following).

Let us remember again, as already said in Chapter 16, that the concept of relativistic mass refers to $\xi(\lvert\mathbf{v}\rvert)$, which is the amplification factor - due to the kinetic state - of the rest emp of a system. Furthermore, the relativistic mass concept has been used successfully - and for many years - in the design of large particle accelerators. That is, the correction for relativistic mass must be applied, even though the standard formula is certainly wrong. However, since it is a first order correction - i.e. with only the first term of a Taylor series expansion - it is not possible to appreciate the mistake. Only in the case of very high "*degraded*" emp velocities could you appreciate the need to use a correct value of $\xi(|\mathbf{v}|)$.

An additional warning deals with the statement that every type of emp interacts through "gravitation" with every other type of emp. This is not always true because, according to what is stated in Chapter 15, there is no gravitational interaction between photon and photon, while there is gravitational interaction between photon, which is "*undegraded*" emp, and the Sun or a star, which are "*degraded*" emp. In fact, it must be considered that rotational kinetic energy has a gravitational interaction with other forms of energy. That is, a rational answer can be given only after considering the true nature of gravitation.

In any case, the distinction between "*undegraded*" and "*degraded*" emp is a factual reality expressed in a "simply" and "beautifully" way.

Compton scattering

For clarity, let us consider the case of Compton scattering (see Chapter 15). For the moment, we consider only the translational motion and always assume that $\beta = |\mathbf{v}|/c < 1$. Let's consider an ideal experiment, in which a photon of energy $h\nu$ and momentum (total intensity) $h\nu/c$ collides with a particle of rest mass (emp) E_0. Before the collision, the particle moves at a velocity $\mathbf{v}_1$ and after the collision, the photon is totally absorbed by the particle, while the particle moves at a velocity $\mathbf{v}_2$. We write as follows the balance, respectively, of energy and of momentum (multiplied by c, for reasons of formal uniformity)

$$E_0\,\xi[(\mathbf{v}_2/c)] - E_0\,\xi[(\mathbf{v}_1/c)] = h\nu \qquad (21.12a)$$

$$E_0\,\xi[(\mathbf{v}_2/c)]\,(\mathbf{v}_2/c) - E_0\,\xi[(\mathbf{v}_1/c)]\,(\mathbf{v}_1/c) = h\nu \qquad (21.12b)$$

The first line, i.e., (21.12a), says that the emp of the particle with rest mass (emp) E_0 after scattering, minus the emp of the same particle before scattering, is equal to the emp of the incident photon.

The second line, i.e., (21.12b), is the equivalent formula referring to momentum.

Indeed, note that (21.12b) represents three equations, everyone with reference to a Cartesian axis in 3D Euclidean space, as $\mathbf{v}$ is a vector with three components.

Express the equality between the first members of (21.12a) and (21.12b), both of which are equal to $h\nu$. Therefore, it must be

$$\begin{aligned}&\xi[(\mathbf{v}_2/c)] - \xi[(\mathbf{v}_1/c)] \\ &= \xi[(\mathbf{v}_2/c)]\,(\mathbf{v}_2/c) - \xi[(\mathbf{v}_1/c)]\,(\mathbf{v}_1/c)\end{aligned} \tag{21.13}$$

Let's consider the case where the particle is at rest before the scattering, i.e., $\mathbf{v}_1 = 0$. For simplicity, and without loss of generality, let's assume that we refer to the absolute reference system, i.e., to æther, and this will be called the reference system of A (Alice). Hence, (21.12a), (21.12b) and (21.13) become

$$E_0\,\xi[(\mathbf{v}_2/c)] - E_0\,\xi[(0)] = h\nu \tag{21.14a}$$

$$E_0\,\xi[(\mathbf{v}_2/c)]\,(\mathbf{v}_2/c) = h\nu \tag{21.14b}$$

$$\xi[(\mathbf{v}_2/c)] - \xi[(0)] = \xi[(\mathbf{v}_2/c)]\,(\mathbf{v}_2/c) \tag{21.14c}$$

Solve (21.14c) to obtain

$$\xi[(\mathbf{v}_2/c)] = \frac{\xi[(0)]}{1-(\mathbf{v}_2/c)} \tag{21.15}$$

which expresses the contraction or expansion of units after the photon diffusion with the particle.

Now let's consider the identical phenomenon observed in the reference system of B (Bob) where $\mathbf{v}_2 = 0$. That is, the reference frame of A, as seen from B, moves relative to the reference frame of B at a velocity $-\mathbf{v}_2$, where the vector $-\mathbf{v}_2$ is oriented in the negative $\boldsymbol{x}$-axis direction. Furthermore, in the reference system of B, the particle before the collision moves with velocity $\mathbf{v}_1 = -\mathbf{v}_2$. From (21.12a), (21.12b) and (21.13) we get

$$E_0\,\xi[(0)] - E_0\,\xi[(\mathbf{v}_1/c)] = h\nu \tag{21.16}$$

$$-E_0\,\xi[(\mathbf{v}_1/c)]\,(\mathbf{v}_1/c) = h\nu \tag{21.17}$$

$$\xi[(0)] - \xi[(\mathbf{v}_1/c)] = -\xi[(\mathbf{v}_1/c)]\,(\mathbf{v}_1/c) \tag{21.18}$$

Solve (21.18) and get

$$\xi[(\mathbf{v}_1/c)] = \frac{\xi[(0)]}{1-(\mathbf{v}_1/c)} \tag{21.19}$$

which expresses the stretching or contraction of units after the scattering of the photon with the particle, as inferred from B, and where, as already mentioned, by definition it is $\mathbf{v}_1 = -\mathbf{v}_2$.

Consider that, when $\mathbf{v} = 0$, it must be $\xi_{tr}(|(\mathbf{v}/c)|) = \xi_{tr}(0) = 1$. Furthermore $\xi_{tr}(|(\mathbf{v}/c)|)$ must satisfy the *ABC theorem*, just as is required for χ, i.e., for "*realism*" (see Chapter 3) it must be

$$\xi[(|\mathbf{v}|/c)]\,\xi[(-|\mathrm{v}|/c)] = 1 \tag{21.20}$$

In contrast, it is found

$$\frac{1-(-\mathbf{v}_2/c)}{1-(\mathbf{v}_2/c)} \neq 1 \tag{21.21}$$

That is, the *ABC theorem* is not satisfied. Therefore, the entire aforementioned formulation has an internal inconsistency.

In fact, as we will see, (21.15), (21.19) and (21.21) express only approximately the correct behavior, as these formulas represent only the first order term of a Taylor expansion with respect to $\beta = (|\mathbf{v}| \,/\, c)$

The Taylor expansion[2] is a well-established algorithm (i.e., a computational tool) of mathematical analysis. It expresses any given continuous function as a sum of infinite terms (called a "series") where every subsequent term expresses a correction of progressively decreasing amplitude until obtaining - at infinity, with the last hypothetical addendum - the true value of the continuous function.

Therefore, it is necessary to define a physical formulation of the dynamics that is more appropriate, in order to satisfy the *ABC theorem*. However, we want to maintain the axiom of the conservation of energy (i.e., of emp) and momentum. But we need to improve the definition (21.1). Therefore, instead of $\boldsymbol{p}$ a new quantity can be introduced, i.e., the "*generalized momentum*" $\vec{P}$ defined as

$$\begin{aligned} c\,\vec{P} \equiv Ef\left(\frac{|\mathbf{v}|}{c}\right)\left(\frac{\mathbf{v}}{c^2}\right) \equiv mc^2 f\left(\frac{|\mathbf{v}|}{c}\right)\left(\frac{\mathbf{v}}{c^2}\right) \\ \equiv m_0c^2\xi\left(\frac{|\mathbf{v}|}{c}\right)\left(\frac{|\mathbf{v}|}{c}\right)\left(\frac{\mathbf{v}}{c^2}\right) \end{aligned} \tag{21.22}$$

where $f(|\mathbf{v}|/c)\;[\mathbf{v}/c^2]$ substitutes the simple factor $[\mathbf{v}/c^2]$ in (21.1).

However, the weight function $f(|\mathbf{v}|/c)$ must be suitably defined. That is, except for the factor $1/c^2$, the generalized momentum resembles a (weighted) flux of E. In the case of a photon, (21.22) becomes

$$c\,\vec{P} = E\;f(|\boldsymbol{c}|/c)\;[\boldsymbol{c}/c] \tag{21.23}$$

where c is the speed of the photon in 3D space, as observed in α world or γ world. Therefore, in the case of a photon, it must be $f(1) = 1$ consistently with what was said before.

In other words, we must assume $f(|\boldsymbol{c}|/c) \equiv f(1) \equiv 1$ if we want the present reformulation to be in accordance with previous standards of physics.

It is shown below that this choice implies no loss of generality. Furthermore, note that, if we want to maintain the rule according to which, in world

---

[2] Brook Taylor (1685–1731) British mathematician.

α or in world γ, respectively, neither a massive particle nor an antimassive particle can be accelerated to the speed of light c, it must be

$$\lim_{\mathrm{v} \to \mathrm{c}} \xi(|\mathbf{v}|/c) f(|\mathbf{v}|/c) \ = \infty \qquad (21.24)$$

i.e., the divergence at infinity is caused by $\xi(|\mathbf{v}|/c)$ which is actually defined as above by (21.11), while $f(|\mathbf{v}|/c)$ is just a corrective factor such that $f(1) = 1$.

But we do not want to delve here into the technicalities of the mathematical analysis of continuous functions, and we only show the final result. The generalized form of momentum, i.e., momentum, is

$$c\,\vec{P} = E\ f(|\mathbf{v}|/c)\ [\mathbf{v}/c] \equiv -\,E\left\{+\sum_{p=1}^{\infty} a_{2p+1}\ (|\mathbf{v}|/c)^{2p+1}\right\}$$

where $$\sum_{p=1}^{\infty} a_{2p+1} \ = \ -2 \qquad (21.25)$$

where $\{a_j\}$ ($j \ = \ 1, 2, \ldots$) denote suitable unknown numerical coefficients.

Note that the coefficient $\xi(|\mathbf{v}|/c) f(|\mathbf{v}|/c)$ is the equivalent of the Einstein coefficient $1/\sqrt{1-\beta^2}$, where as usual $\beta = \mathrm{v}/c$. However, the Taylor expansion of the Einstein coefficient contains only *even* powers of $\beta = \mathrm{v}/c$, while the analogous corresponding - and correct - formula in (21.25) contains only *odd* powers of β.

Therefore, Einstein's coefficient $1/\sqrt{1-\beta^2}$ *cannot fit the ABC theorem in any way.*

Thus, according to the axiomatic formulation discussed here, *the standard relativistic formula commonly used for the Doppler effect is certainly incorrect.*

Note that (21.25), together with the aforementioned conservation laws, is the formal substitute for Newton's second law of dynamics.

If we perform a Taylor expansion of $\xi[(|\mathbf{v}|/c)] f(|\mathbf{v}|/c)$, and truncate the expansion to first order, we find the classical theory of relativity (both special and general).

Indeed the coefficients $\{a_{2p+1}\}$ ($p = 0, 1, 2, \ldots$) are unknown. The case of classical relativity is found when all coefficients $\{a_{2p+1}\}$ vanish except the first.

However, no physical reason seems to suggest such an inference. In any case, the formula is certainly non-linear, as the truncated expansion to first order cannot satisfy the *ABC theorem*. Therefore, non-linearity is a constraint that *must* always be satisfied.

Note that, if $\mathbf{v}$ is known, it is possible to calculate two quantities, i.e., both the generalized momentum using (21.25), and also $\xi(|\mathbf{v}|/c)$ using a formula not shown here. Therefore, if the rest mass of the granule is known, the total emp content can be evaluated, i.e., the role of kinetic energy and rotation energy. Conversely, it is not possible to provide a simple inversion formula suitable for calculating $\xi(|\mathbf{v}|/c)$ when both the total emp and the generalized momentum are known.

That is, unlike the rigorous axiomatic value in absolute terms of Newton's formulation, the use of the generalized moment represents a mediated, "statistical" treatment aimed solely at seeking a connection with the classical formulation according to Newton. In reality, a strictly rigorous formulation must refer to Feynman graphs, to volleys that are "*lame*" or not, etc.

A last question remains to be clarified. How would the formulation develop if - as mentioned above - instead of the *multiplicative* criterion, the *additive* criterion is adopted?

We don't dwell on formal mathematical technicalities. Everything is traced back to an appropriate mathematical object ("*operator*") denoted by $\mathcal{O}(\vec{\mathrm{v}})$ whose most general expression must satisfy the required constraints. Therefore, this "operator" is obtained by representing $\mathcal{O}(\vec{\mathrm{v}})$ with a function, or with a Taylor series expansion, which is shown to be

$$\mathcal{O}\left[\left(\frac{\mathbf{v}}{c}\right)\right] = \sum_{k=0}^{\infty} b_{2k+1} \left(\frac{\mathrm{v}_{\parallel}}{c}\right)^{2k+1} \tag{21.26}$$

where $\{b_{2k+1}\}$ is a given set of coefficients, to be compared with the $\{a_{2p+1}\}$ coefficients appearing in (21.25). The coefficients $\{b_{2k+1}\}$ are called with a different symbol to avoid confusion with the coefficients $\{a_{2p+1}\}$. The symbol $\mathrm{v}_{\parallel}^{k}$ denotes the kth power of the component of $\mathbf{v}$ along the source-observer line of sight. Note that $\mathcal{O}(0) = 0$. Additionally, note that only odd powers of $\mathrm{v}_{\parallel}$ are considered to satisfy the requirements of the *ABC theorem*.

We omit subtle consequences on the Doppler effect formula (see Chapter 11).

In reality, both *multiplicative* and *additive* logic can be applied, but not both. The phenomenon, the physics, are the same, only the mathematical formalism changes, the "language" to describe them. The concern is therefore to imagine an experiment designed to evaluate the coefficients, $\{a_{2p+1}\}$ or $\{b_{2k+1}\}$, and to choose whether natural reality adapts to one of the two choices (not both). Note that, in any case, both $\{a_{2p+1}\}$ and $\{b_{2k+1}\}$ can always be evaluated formally. The choice is therefore based on the relatively faster convergence of the Taylor series expansion and how the mathematical formalism best fits the observations.

We emphasize how the estimate of the coefficients $\{a_{2p+1}\}$ or $\{b_{2k+1}\}$ is based on concrete experimental observations - inherent to the Pioneer anomaly or other experiments available in the future. Nothing is abstract or based on mere speculation. Rather, it is a matter of evaluating the minimum required for the observed data in order to avoid unfortunate paradoxes.

With this, we have completed the brief discussion of the connection between the axiomatic approach considered here and the classical formulation according to Newton/Maxwell/Einstein. Now there remain some very interesting collateral aspects to consider.

Chapter 22

Superluminal Phenomena
The Doppler Effect
Antimatter as a Superluminal Phenomenon?
The Law of Conservation
of the Time-Antitime Flux

"Every theory is beautiful. Observations - how much are they objective? There can be several different interpretations of observations. I do not like that experiment decides about theory. Theory is freedom of thinking. Freedom."

Zbigniew Oziewicz (1941-2020)
[private communication, 2011; mathematician and Polish freedom fighter, member of *Solidarność Walcząca*)]

....

Preamble

Our journey becomes more and more daring. Are we really sure that nothing can move at a speed greater than light speed? Let's try to ask ourselves the question of whether "*superluminal*" phenomena can exist, i.e., phenomena that propagate at a speed greater than light speed.

Nothing denies the fact that in our world α and in the specular world γ every massive or antimassive object (i.e., composed of mass or antimass) moves at a speed lower than light speed, as it is composed of "*degraded*" emp. However, no reason can exclude *a priori* that some phenomena that we observe - or not - are actually phenomena from other worlds where the speed of motion is greater than that of light.

Let's see how matters are, and we will find some very interesting indications dealing with our world as well, in addition to other possible worlds. The surprises of our logical journey just seem to never end!

Anticipating the final result, we will see how superluminal phenomena may well exist, although - as a standard - some galaxies are interpreted according to the big-bang logic, while they can also be considered imploding towards the Earth at a velocity $|\mathbf{v}| > |c|$. Furthermore, men can well consider the observations both in terms of the logic of expansion, i.e., of the big-bang, and in terms of an implosion towards us at a speed $|\mathbf{v}| > |c|$.

This follows from a property of the topology (i.e., the general geometric properties) of the worlds α, β and γ (see Chapter 23). We will discuss some details regarding materialization and annihilation of particles/antiparticles, depending on the conservation of time/antitime, as strictly required by the coexistence of conjugate complex twin worlds α and γ, respectively.

The interaction between different objects occurs through the *SEA* mechanism ("*Steady Exhalation and Annihilation*", see Chapter 14 and Chapter 23). Owing to a property endemic to the β world, every emp content eventually materializes into a pair of "*elementary granules*" of opposite polarity, which are composed of "*degraded*" emp in the α and γ world, respectively.

Dynamics and antidynamics occur through an interplay of "*double helices*", "*single helices*" and "*boosters*". The history of the Universe thus proceeds almost paradoxically - cyclically in a timeless world. Galaxies composed of matter in the α world, and antimatter galaxies in the γ world, are generated symmetrically, and then annihilate cyclically, without stopping in the timeless β world.

Finally, we will discuss the nature of the "*thermodynamic*" approach, necessary whenever a large set of monads must be considered simultaneously (see Chapter 18).

For the time being, we will better analyze a concept that is crucial to our discussion, namely, the relationship between the Doppler effect and superluminal phenomena.

Introduction

We must consider several innovations with respect to the standard approach of today's theoretical physics. In any case, a fundamental criterion is Ockham's razor (see Chapter 14), i.e., the axiomatics must be as "simple" as possible, and any excessive complication must be avoided, unless it appears strictly necessary.

In other words, we must always draw inspiration from Franciscan humility in the face of Mother Nature, without worrying about trying to explain "everything". A "beautiful" theory that disagrees with observations is better than a "not beautiful" theory that satisfies every observational constraint, as emphasized by Dirac in 1963. However, it is unfortunate that the "everything" cannot be deduced from a "simple" or linear scheme. Even when you try to simplify as much as possible, you can't avoid some topics that seem really intricate to us.

A key issue is related to the Doppler effect. The standard relativistic formula is certainly incorrect, although it is commonly applied in astrophysics.

The correct formula can only be derived by means of the so-called *Pioneer* anomaly, which, at the moment, seems to be the only available experiment with which it is possible to get rid of the related unfortunate impasse (see Chapter 16).

In general, we will find that the search for a "beautiful" and self-consistent theory leads to the consideration of the possible existence of superluminal phenomena, although humans are perhaps prohibited from detecting them.

Indeed, a double consequence can apparently be foreseen. On the one hand, we gain a better understanding of the mechanism governing matter/antimatter, "annihilation" and "materialization". On the other hand, one can imagine some unexpected conclusions concerning the so-called "*red-shift*", the "expansion of the Universe", the "big-bang", and associated cosmological phenomena.

The "*red-shift*", i.e., the shift towards red, refers to all spectral lines of a given chemical element, which are recognized in the spectrum of a celestial body (galaxy or other). All these lines denote a shift towards longer wavelengths, i.e., towards red. This phenomenon is interpreted, owing to the Doppler effect, as evidence - and measure - of the light source moving away from the Earth. The observation of the *red-shift* is the basis for hypothesizing the expansion of the Universe, an expansion that would have begun at a certain initial instant as a great explosion, literally translated as "big-bang".

However, as a necessary premise, since the relativistic Doppler formula is certainly wrong, all observed astrophysical implications must be critically reevaluated, as soon as the correct formula becomes available, in a way consistent with the discussion given in Chapter 16.

It should thus be discovered that the Universe is composed of both expanding and imploding emp, although humans have a different perception of expanding or imploding galaxies, depending on the superluminal emp.

In this way, you can obtain a "beautiful" theory. Several cosmological issues are thus reconsidered. That is, we can imagine some innovative perspectives, which concern, e.g., the increase in the entropy of the Universe (see Chapter 4), or the (seemingly paradoxical) aging of the Universe, or the future evolution of the Universe, etc.

In any case, whether correct or not, the formulation presented here should be considered as the basis for discussion. In fact, we cannot free ourselves from paradoxes if we do not question our certainties, paradigms, and current beliefs that are "generally agreed upon" and considered "established".

Doppler Effect and superluminal phenomena

The classic formula for the Doppler effect is based on the simple logic of Fig. 11.2. If the *additive* criterion is adopted (see Chapter 21), the classical formula is coherent with the *ABC theorem*. On the contrary, if the choice is based on the better well-known *multiplicative* criterion, the classical formula is not consistent with the *ABC theorem*. However, the standard and well-known relativistic formula - which is admittedly wrong - is based on the multiplicative criterion and is inconsistent with the *ABC theorem*.

Therefore, all applications to astrophysics (*"red-shift"*, big-bang, cosmology, etc.) must be critically reconsidered, after the evaluation of the correct set of coefficients, both $\{a_{2p+1}\}$ and $\{b_{2k+1}\}$, which enter the definition of the correct formula (see Chapter 21).

Therefore, the more general analytical form for $\chi[(v/c)]$ must be considered. The primary physical and logical basis is based on *(i)* the principle of *"realism"* and the principle of relativity (see Chapter 3), *(ii)* the principle of constant value of c, and *(iii)* considerations of symmetry.

Furthermore, it must be considered that - in principle - the entire Universe evolves as a single whole (although separately in the α world and the γ world, which are complex conjugates of each other). This uniqueness of a natural system is synonymous with *"irreversibility"*, including the explanation of the univocal definition of the direction of time and antitime, respectively in the world α and in the world γ (see Chapter 9).

On the contrary, if we take into account the need for "simplicity" required by the human mind and if we separate the Universe into subsystems, this approximation is equivalent to the loss *ab origine* of the perception of *irreversibility*. We then must deal with natural laws, which all seem necessarily reversible.

It seems embarrassing, if not impossible, to imagine any further properties that we might be able to learn about, unless we are able to obtain further observational data.

Furthermore, one must consider that every experiment, of the entire enormous set of available observational tests of relativity, is generally concerned with small ratios $\beta = |\mathrm{v}/c|$. Hence, only a mere first-order linear approximation can be checked. In fact, every experiment based only on small values of the ratio $\beta = |\mathrm{v}/c|$ is not suitable for testing the more general nonlinear law. At the moment, the nonlinear test seems to be useful for this purpose is only through the *Pioneer* anomaly (see Chapter 16).

In other words, there is no contradiction between our journey and the set of all previous available experiments that are claimed to confirm relativity.

There is only a need to go beyond the linear or first-order approximation, which is the typical realm of classical special relativity.

Therefore, concerning laboratory experiments, in general, this conclusion has no relevant consequences, while it can have a relevant implication only when astrophysical phenomena are considered, as shown by stupendous images observed by telescopes outside the Earth's atmosphere (see a few mentions in Chapter 26).

In summary, all astronomical indications derived from the Doppler effect (including, e.g., evidence on the expanding Universe) require a critical and substantial analysis, and/or possible reinterpretation.

We consider the case of a photon as a volley of miniphotons (Fig. 10.3), i.e., $a_1 = 1$ and all the others $a_{2p+1} = 0$ (for every $p \neq 0$) - and, in a similar way, if we adopt the additive criterion instead of the multiplicative one, we consider the case $b_1 = 1$ while all the others $b_{2p+1} = 0$.

Therefore, there are two possibilities. One possibility is that the model of the photon as a volley is a simplified "mnemonic" rule – analogous, e.g., to the Bohr atom to symbolize the Schrödinger equation etc. (see Chapter 15) – or, a second possibility is that every other particle can also be described symbolically in terms of a volley of suitable miniquanta, similar but different from miniphotons.

In general, miniquanta - both photons and others - are therefore to be called more generically "*elementary granules*".

On the other hand, a possible, or perhaps very probable, distinction between different particles depends only on the different length of the volley of miniphotons, with no need to appeal to different types of "elementary granules" (see below).

Note, however, that when we rely on the miniphoton/"*elementary granule*" hypothesis, we implicitly make a reference to some hidden parameters. That is, in strict principle, owing to the empirical constraint, we can observe no hidden parameters (see Chapter 1). The volley model should be considered only as a "mnemonic" rule, or a "simple" or "beautiful" way to represent some real physical information, although of an unknown nature.

In other words, we describe the system in terms of the "*virtues*" of the monads that we can only observe from the outside (see Chapter 4), while we are not capable of evaluating the final, real, and intrinsic physical detail of every "*virtue*". However, upon careful analysis, the study of the Doppler effect leads to some unexpected conclusions.

Let's begin and distinguish three different types of Doppler effect:

(i) the *emission* Doppler effect, i.e., associated with the motion of the photon source;

(ii) the Doppler effect generated by the movement of a *mirror*, which reflects the photons originating from a source;

(iii) the *environmental* Doppler effect, i.e., associated with a variation in the frequency ν of the photon, caused by some environmental condition (e.g., gravitation) that influences the photon.

The ambient Doppler effect can act directly on the light source, or even on the photon crossing spacetime. Fig. 11.2 is the symbolic representation of the Doppler effect of emission.

But the same element can be represented symbolically by a different and "simple" model (Fig. 22.1). In fact, let's consider a conveyor belt which – according to the information available to the observer – is known to have no limits to the right or to the left (i.e., it is not infinite, even if we don't know where it ends). Let A (Alice identified with us) stand at point A and let A observe some seeds which are regularly dropped on the tape. The tape - at least as far as A is aware - seems to run at a regular speed.

Seeds are dropped onto the conveyor belt by a sower S, which is located at a point along the conveyor belt, although its position is unknown to A. The conveyor belt moves at light speed c and the seeds are miniphotons of a photon. The sower drops a seed at every time interval $\tau \equiv 1/(2\nu)$ and at a regular rate (at least as far as A can estimate or presume to evaluate, according to his clock).

In Fig. 22.1a the sower S is in a fixed position relative to A. Thus, A observes the seeds positioned on the conveyor belt at a respective regular distance $\lambda/2 \equiv c/(2\nu)$. Let us instead suppose that S moves towards A at a speed $0 < |\mathbf{v}| < c$ (Fig. 22.1b). Observer A detects the seeds at a regular reciprocal distance $\lambda' \equiv \lambda[1 - \mathrm{v}/c]$. Instead, if S moves away from A (Fig. 22.1c) with a velocity $\mathbf{v} < 0$ and $|\mathbf{v}| < c$, Alice (A) detects the seeds at a regular reciprocal distance $\lambda' \equiv \lambda[1 + |\mathbf{v}|/c] \equiv \lambda[1 - \mathbf{v}/c]$. Such effect is summarized in Fig. 22.2.

The same result applies to both the "*emission*" and the "*mirror*" Doppler effects, because the effect is the same whether the sower S takes the seed from a basket (*emission* case) or whether he receive one seed at a time (*mirror* case) and then he put it on the conveyor belt, almost like a tennis racket hitting and sending the ball in the chosen direction.

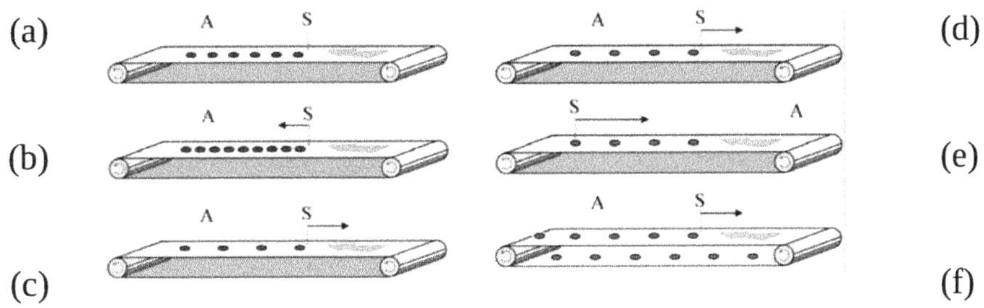

Fig. 22.1. The structure of a photon is shown symbolically in Fig. 22.1a, 22.1b, 22.1c (from top to bottom) [*left column*] and in Fig. 22.1d, 22.1e, 22.1f [*right column*]. A conveyor belt is almost infinite in length. The observer (Alice) is at A, while a sower is at S. However, A does not know the position of S. The tape moves at a uniform speed c. The sower drops one seed at a time onto the belt at regular and constant time intervals (the term "constant" refers to A's clock). The position of S with respect to A is constant in Fig. 22.1a. Conversely, in Fig. 22.1b, S moves towards A with a given speed $\mathbf{v}$ and, in Fig. 22.1c, S moves away from A with a given speed $\mathbf{v}$. Fig. 22.1d is the same as Fig. 22.1c. In Fig. 22.1e, S moves rightward at a speed that is twice the speed of the conveyor belt. In the case of Fig. 22.1e, A observes the same effect she can observe in Fig. 22.1c. However, in Fig. 22.1c, S escapes at a velocity $\mathbf{v} > c$ outward from A, while, in Fig. 22.1e, S approaches A with a velocity $\mathbf{v} > 2c$. In the case of Fig. 22.1e, A first detects the miniphoton that was released relatively later by S. Thus, according to A, there is a reversal of the time sequence, i.e., apparently a reversal of the arrow of time. This is the world of superluminal phenomenon. Fig. 22.1f shows that, in certain respects, the same effect can be described by a different and simpler model, in which the conveyor belt is transparent, through which A can see the miniphotons that are attached to the conveyor belt and which flow on the underside of the conveyor belt, i.e., the miniphotons seem to move along the opposite arrow of time. This is a different way of saying that A can believe *(i)* that she can observe phenomena with a conventional arrow of time, even though they are superluminal phenomena, or, equivalently, *(ii)* that she can observe the same events with a reversed arrow of time, even though with speed $|\mathbf{v}| < c$. That is, A can observe the effect of miniphotons moving on both sides of the conveyor belt, i.e., along the arrow of time or along the arrow of antitime, depending on whether A prefers to consider an antiworld with an inverted arrow of time and emp negative, or, rather, she prefers to maintain the direction of the arrow of time, although she admits the existence of superluminal phenomena. This mechanism is reminiscent of the concept of "rubber band" (see Chapter 14 and Chapter 15), or - even better – of the "double helix" shown in Fig. 6.2 and 10.4.

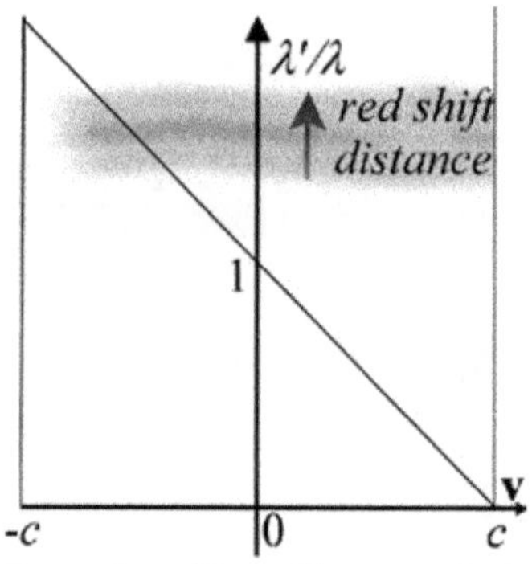

Fig. 22.2. Result of the Doppler effect (from *emission* or from moving *mirror*) in the case where everything in the Universe must move at a certain speed $|\mathbf{v}| \leq c$. The symbols λ and λ' indicate the wavelength, unaffected or affected by the Doppler effect, respectively. The linear progression expresses the first order approximation of the Doppler effect, which does not satisfy Einstein's *ABC theorem*. The correct and not approximate trend is perhaps, although probably, a slight variation of this approximate linear trend.

Note that, consistent with the observational evidence provided by Michelson-Morley experiments (and similar experiments; see Chapter 17), the logic of Fig. 11.2 and Fig. 22.1 is as in the case where a light emitter (i.e., the sower S) releases the photon at a speed *c*, and immediately afterwards "forgets" about it. In fact, the photon is a "*virtue*" of the monad's atom, and the photon, after emission, moves through 3D space and forgets the atom that emitted it.

The cases of superluminal phenomena (Fig. 22.1d, e, f) are discussed further below. In both cases of Fig. 22.1b or 22.1c, the motion of S is such as to modify the distance between the successive miniphotons within the photon volley, i.e., on the conveyor belt. This mechanism is based only on experimental evidence, i.e., it is not a question of any theorem or result that derives from some fundamental law of physics or other argument. Rather, it is an inherent logic in the chosen "simple" model – and it seems to be strictly consistent with observations.

The discussion presented here is only intended to highlight some implications that are thus deduced. Before proceeding, however, let us consider the case of the *mirror* Doppler effect in detail. It is essential to first consider a physical definition of "*mirror*".

The concept of a *mirror* is an abstraction - just like different concepts in geometry (such as a point, a line, a plane, a fractal, ...), or in mathematics (such as perfect randomness, infinite or infinitesimal quantity, etc.), or as any concept that refers to an ideal physical behavior (e.g., ideal rigidity or elasticity, or ideal plasticity, or a "Newtonian" fluid, or a perfect conductor, or an ideal superconductor, or a perfect insulator, or a massive object considered point-like, or a non-deformable "solid" object, or a state of strictly null absolute temperature, etc.).

In principle, the abstract idea of a "*mirror*" implies *(i)* an ideal infinitely thin surface, and *(ii)* an ideal "reflective" surface. That is, we arbitrarily introduce a new hypothesis and assert that a natural system behaves like a "mirror" if it reflects something in a similar way to the reflection law of elementary optics. That is, the direction of the reflected image must be symmetrical with respect to the perpendicular to the reflecting surface drawn at the "point of incidence" of the incident physical image. Physical reality can provide several examples of "mirrors", which refer to physically very different scales.

For instance, consider the case of classical physics. Consider a region D of space, as a sort of "surface" with physical thickness $k \neq 0$, i.e., consider a given layer D (of any shape, i.e., not necessarily a plane), which has an "external" side and an "internal" one, and is composed of a material that has an electrical conductivity σ. Consider a plane and homogeneous electromagnetic wave that impacts D from outside. Owing to electromagnetic induction, some electric currents are generated within D (see Chapter 14 and Chapter 15 and *passim* elsewhere). The incident electromagnetic signal is damped as it penetrates deeper into the medium, according to an exponential damping characterized by the so-called "penetration depth", which is a function of the frequency ν (i.e., a higher ν has a shallower penetration depth). This effect is also known as "*skin effect*". This damping precisely expresses the fact that the material of which the D layer is composed tries to reduce or cancel the disturbance due to the incident electromagnetic wave.

The electromagnetic field measured by an observer who is located externally to D, is similar to the case of a reflection by D of the incident electromagnetic signal.

Finally, carry out a limit process, i.e., let $\sigma \to \infty$ and $k \to 0$, and in this way, with a simple abstraction, an ideal "mirror" is defined, of infinitesimal thickness, which imitates the behavior of the system - according to our abstract and simple pattern - when observed from the outside.

In contrast, when the reflection process applies to a single photon, we must refer to the interaction between a photon and another object (a particle or a photon). That is to say, the same "macroscopic" concept of "mirror" can hardly be applied, as the process rather concerns the interaction between individual photons/particles. That is, the details concerning the individual photons/particles must be considered separately.

Another similar example concerns the solar wind, which is generally conceived as an approximately ionized medium with $\sigma \to \infty$. For the sake of brevity we will not go into details here, which would require some non-essential technicalities. Let us just remember that if we assume that this property is strictly valid within the solar wind for every frequency, it should be strictly impossible for us to observe every emission released by the Sun, including every ray of light or every radio emission, etc. That is, we shouldn't even see the Sun, just as we can't see its interior, which is completely shielded by its highly conductive outer layers. In other words, the Sun would paradoxically be a perfectly impenetrable, dark, and invisible monad for us.

In reality, we hypothesize that the solar wind acts like a perfect "mirror" – i.e., just like an approximately infinitely thin and omnipresent reflecting surface - although this approximation does not apply to visible light or radio waves, etc. Therefore, the solar wind is a "perfect mirror", but only with reference to a given frequency band of electromagnetic radiation.

As far as our discussion is concerned, we ask whether there is a "mirror" in the Universe (i.e., within our "big-bang monad"). In general, the limit process for $\sigma \to \infty$ and $k \to 0$ is only a "simple" abstraction according to the needs of simplicity required by the human mind, although this abstraction is *per se* nonphysical.

On the other hand, there is no need that an infinitesimally thin reflecting surface ($k \to 0$) and with $\sigma \to \infty$ exists in the Universe. We can well consider physical objects, even having a huge k and a limited σ - and which, when observed from Earth, behave exactly the way of a mirror as they reflect photons. That is, as far as our final observations are concerned, such a type of physical structure seems to behave like an ideal "mirror".

Summing up, we define a straight line by assimilating it to a ray of light. In much the same way, we consider a reflected ray to be "perfect" and similar to a ray of light of normal emission. In other words, instead of dealing with the abstract concept of a straight line, we deal with the abstract concept of a straight line interrupted by a surface, and "reflected" back from a surface, while the concept of a surface is an additional abstraction.

That is, "*mirror*" is a conventional way of describing the curvature of the light ray path, much like that of a waveguide, or a lens, etc., while the physics of the photon is implicitly assumed to be completely unaltered, and in particular it is assumed that the mutual distance between the successive miniphotons along the volley is not affected by reflection, if the "mirror" is not moving with respect to us.

Owing to a physical reason, the energy $h\nu$ of the photon is related to the emission process from the atom-monad that is the source of the photon. Furthermore, a change in the frequency ν is simply caused by the Doppler effect, associated with the relative motion of source and observer, or mirror and observer, but the change in frequency ν does not modify the total energy transported by the miniphoton volley.

For the time being, let us therefore forget the details of every celestial "mirror" and consider how - compared to an emission Doppler effect - a miniphoton volley should react to reflection in terms of a mirror Doppler effect.

As for the $|\mathbf{v}| < c$ case, the argument mentioned above applies to both types of Doppler effect. Let us consider, however, also the hypothetical possibility that $|\mathbf{v}| \geq c$ (Fig. 22.1d, e, f). Consider in great detail the caption of Fig. 22.1d, e, f which is fundamental to understanding the whole discussion of superluminal phenomena. This is in contrast to the axiom of the maximum "asymptotic" speed c. However, this hypothesized and unprecedented possibility leads to some unexpected consequences that, perhaps, have some important correspondence in real observations.

Our argument is based on the ratio between inertial mass and rest mass, which is divergent when $|\mathbf{v}| \to c$. This conclusion follows from the fact that ξ and χ coincide (see Chapter 21), and when the source moves at a velocity $|\mathbf{v}| \equiv c$, the Doppler effect on a photon implies the paradoxical conclusion that $\lambda \to 0$ or $\nu \to \infty$.

Consider, however, the distinction between "*non-degraded*" emp and "*degraded*" emp. See also below. The correct ξ or χ coefficient is given by the generalized form of momentum, whether the multiplicative criterion is adopted or the additive criterion (see Chapter 21). In principle, we cannot know *a priori* if one of the two generalized forms implies an asymptote for $\mathbf{v}| = c$ or not. Perhaps there is an asymptote, perhaps not.

As far as we know, no physical reason seems to forbid *a priori* that, somewhere in the Universe, there exists an entity that can possibly propagate at a speed $|\mathbf{v}| > c$, although (perhaps) we are not able to see it (see Chapter 23).

For the moment, we therefore consider this possibility as a provisional working hypothesis. For instance, suppose that all humans are "blind" and imagine the kind of scientific results and knowledge that we humans can achieve. Similarly, we can hypothesize and assume that, in fact, we can be, perhaps, "blind" (or "almost blind") with respect to superluminal phenomena, i.e., with respect to everything that moves (in our reference system) with a wavelength λ' with $|\mathbf{v}| > c$.

In any case, with this working hypothesis, we will find - to our surprise - a "beautiful" mathematical regularity, so that - even if in the end no physical meaning can be associated with superluminal phenomena - the hypothesized existence of superluminal phenomena can help to understand some characteristics of the observations we can carry out.

If $|\mathbf{v}| > c$, the argument regarding Fig. 22.1a, b, c still holds, based on the identical formulas for λ' etc. This is the case of Fig. 22.1d, e, f, while Fig. 22.2 needs to be transformed into Fig. 22.3.

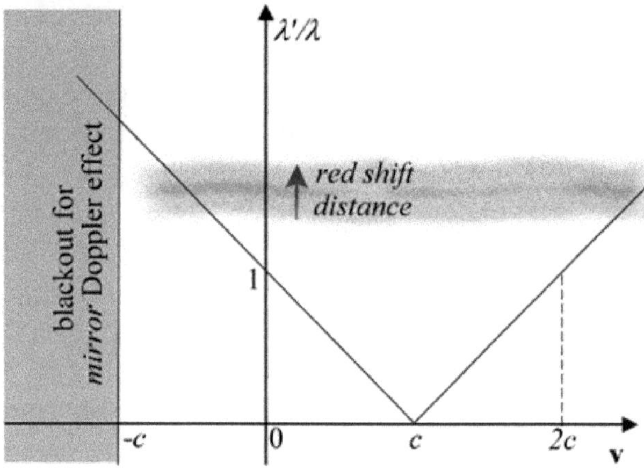

Fig. 22.3. Result of the Doppler effect (*emission* or *mirror*) in the case in which everything in the Universe can move even at speed $|\mathbf{v}| > c$. For $\mathbf{v} < 0$ (i.e., the relative motion of the observers is away from each other) and |v|>c, the mirror Doppler effect is obscured, i.e., humans are unable to detect it. The linear behavior is only the first order approximation (see similar comments in Fig. 22.2). See text.

In fact, consider that, in the case of $-|\mathbf{v}| < -c$, the mirror Doppler effect does not exist - as the mirror should reflect the miniphoton before the time instant of emission from its source, therefore, before the concrete physical availability of the miniphoton, which would be paradoxical in terms of causality, and causality is a yardstick of "*realism*".

Furthermore, consider that a "mirror", which moves at a speed $|\mathbf{v}| \equiv c$ in the same direction as a photon, collects all the miniphotons, almost like a child collecting butterflies with a net. In physical reality, this circumstance should be identified with the limiting case in which the "mirror" should experience a "flash" of light, i.e., an immense instantaneous accumulation of photonic energy in spacetime.

This is the case which in Fig. 22.2 and 22.3 corresponds to the vertex point at $|\mathbf{v}| = c$, where a discontinuity occurs.

On the other hand, in general, when dealing with real phenomena, no discontinuity can occur. Every discontinuity is always the result of some "simplifying" abstraction envisaged by the human mind when using continuous functions.

The same argument applies to the emission Doppler effect. Such a physical occurrence of the effect can occur, e.g., whenever the light source - such as a star - falls towards the Earth at $|\mathbf{v}| \equiv c$. Such a hypothetical catastrophe, in terms of photons (and neglecting the simultaneous gravitational effects (?) on the Earth and the star), should imply for the Earth a huge flash of light. This is the paradox associated with the Doppler effect associated with a source that moves at $|\mathbf{v}| \equiv c$.

That is, this paradox is not physical, and it is not possible in reality. Rather, it is a discontinuity in our plots, associated with the transition to the world β either from the world α or from the world γ. In fact, a discontinuity is never a real physical case. Rather, it always derives from some extrapolation of some "simple" and "beautiful" model that we hypothesize for simplicity's sake. We tend to satisfy the needs of "simplicity" – which are inherent in the human mind – rather than satisfying the needs of objective reality.

Such a discontinuity seems almost like an impenetrable boundary separating the physical realm of all entities with $|\mathbf{v}| < c$ from everything that moves with |v|>c. In other words, we should state that this "boundary" coincides with what moves with $|\mathbf{v}| > c$. That is, this "boundary" is the "*brane*" (abbreviation of "membrane") identified with the β world. Let us remember this definition of "*brane*" for the following. See Fig. 22.4. In fact, when considering a photon, the emp E and momentum can only be considered when a photon interacts with a monad, either in the α world or in the γ world. In the β world the photon is motionless, with no interactions whatsoever. This characteristic is reminiscent of what in mathematics is called an asymptote, which divides the domain of definition of a function that is positive on one side of the asymptote and negative on the other (Fig. 22.4). Therefore, it is impossible to define a value for the function corresponding to the $\boldsymbol{x}$-coordinate of the asymptote. The conclusion is that, as was previously proposed by Dirac a long time ago, it is convenient to state that the observer of the world α assumes that in the world γ is $E < 0$.

In summary, an unexpected – and seemingly paradoxical – complication emerges from Fig. 22.3. In fact, every time we observe a $\lambda' > \lambda$, we have two possible choices, namely, either *(i)* the source of the miniphotons (i.e., the sower S of Fig. 22.1) moves away from A with a suitable $|\mathbf{v}| < c$, or *(ii)* it moves towards A, although at a velocity $|\mathbf{v}| > 2c$.

As far as we can understand - in principle - both choices are equally possible and correct. The implications for astrophysics are discussed in Chapter 24.

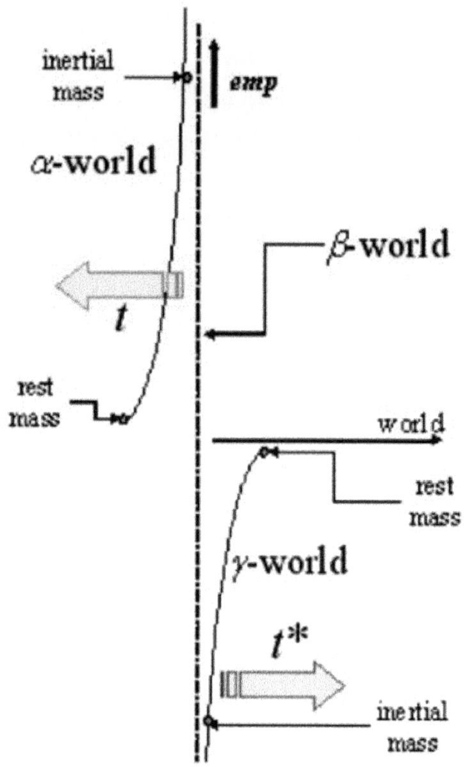

Fig. 22.4. The world β is like an asymptote for the emp E, and is the "*brane*" that divides world α (with time t) and world γ (with complex conjugate antitime t^*, i.e. with inverted arrow of time). Note how on the $\boldsymbol{x}$-axis (i.e., on the horizontal axis) there is a mixture of time and antitime. See text.

However, let us first reconsider the observational evidence concerning "annihilation" or "materialization" of matter-antimatter. That is, let's focus on what happens within a monad, i.e., we consider phenomena that we can only know through the "*virtues*" of the monad. In contrast, every speculation that relates to the interior of the monad must be in terms of hidden parameters. On the other hand, hidden parameters are just an intelligent, arbitrary, and "simple" mnemonic tool (such as the Bohr atom; see Chapter 15). It should be stressed, however, that every symbolic mnemonic tool cannot have any objective meaning, and can be equivalently replaced by any other equivalent type of mnemonic model.

In this same respect, it should be noted that some different symbolisms have been considered in the reformulation of the axiomatics proposed here. A symbolism is the classical Feynman graph. Another symbolism is the representation of miniphoton volleys in Fig. 10.2. A third symbolism is the representation of the "sower conveyor" shown in Fig. 22.1. A fourth symbolism is in terms of a "double helix" shown in Fig. 10.4. The real physical phenomenon is unknown, because it is based on hidden parameters that - due to the empirical constraint - we cannot observe. Thus, everyone of the four symbolic representations is correct, although every symbolism expresses a slightly different detail by comparison.

In this same respect, for completeness and curiosity, we remind that strange phenomena are sometimes reported in the literature such as that of the biphoton, or a photon that divides into two parts. This is conceptually shown in detail with the mechanism shown in Fig. 22.5.

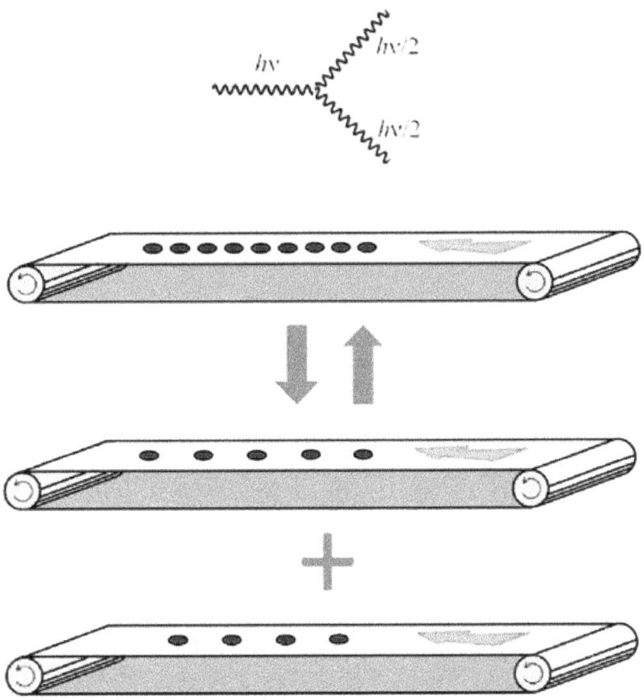

Fig. 22.5. A biphoton arises from the splitting of one photon of emp $h\nu$ into two photons of emp $h\nu/2$ each. The top graph is a standard Feynman graph, which symbolizes the process. Since we can observe the photon from the outside, we cannot know what process occurs inside it that determines the event. The bottom part of the figure shows another symbolic representation equivalent to the Feynman graph, although by means of a different pictorial representation. Note that a photon has an emp which is the sum of the emp of the biphoton. That is, $h\nu = h\nu/2 + h\nu/2$, i.e. the frequencies simply add together, i.e. $\nu\ /\ 2 + \nu\ /\ 2 = \nu$. See text.

We therefore conclude that superluminal phenomena no longer seem as “exotic” a topic as they seemed at the beginning. So let’s explore some other consequences. Of course, our journey never ceases to amaze us...

Antimatter as a superluminal phenomenon? The law of conservation of time-antitime flux

Note that the superluminal argument - extravagant and unusual - is somehow reminiscent of the negative energy states of antimatter, moving from the future to the past and hypothesized by Dirac, Feynman, etc.

In fact, the idea implicit in Fig. 4.1 (i.e., in Fig. 6.5 or 9.1) predicts that, every time a photon contains an excessively large emp, the maximum emp density principle (see Chapter 4) is violated. Thus, the photon must "materialize" into matter-antimatter. That is, starting from the original "*non-degraded*" and timeless world β, two objects are born, respectively, immersed in the two "*degraded*" symmetric complex-conjugated worlds α and γ through a sort of natural "exhalation".

In any case, the conservation laws of emp, of generalized momentum, and of generalized angular momentum must be satisfied (see Chapter 21). Why this happens? For instance, suppose that two photons interact and merge together, e.g., as in a unique biphoton (see Fig. 22.5), but in the reverse way. But how can this occur? We can try to give an answer only through speculation and imagine a "beautiful" theory, although always in agreement with Ockham's razor (see Chapter 14).

One possible answer is that the unique photon, which results from the fusion of two previous photons, splits into two parts that are identical due to conservation laws. However, one half remains in the β world, while the other half decays and becomes "*degraded*" emp split into two parts. A part is in the "*degraded*" world α and moves with velocity $|\mathbf{v}| < c$. The other half enters the γ world, where it is similarly emp "*degraded*".

We can equivalently claim *(i)* that the arrow of time is reversed, or *(ii)* that antimatter enters the superluminal world. That is, the physics is the same for either choice. The choice is left up to the arbitrariness of the observer. In any case, this description is consistent with the materialization/annihilation axiom.

In summary, for symmetry reasons, a further conservation law can be added, as follows.

Axiom - Conservation law: the total flow of time-antitime must always be null

Indeed, the world β is timeless. Thus, in the world β no flow of either time or antitime can occur, because by definition, both time and antitime do not exist. The photon is just "*non-degraded*" emp which, when observed in the α world or in the γ world, by an axiom, can move only at a constant velocity c.

However, when the principle of maximum emp density (see Chapter 4), or some other principle, is violated, the photon must split into two components, one in the conjugated α-world and one in the γ-world, while the total time-antitime flux must always be null. Indeed, the world α is characterized by time with a given direction of the arrow of time, while the world γ is characterized by antitime, with an inverted arrow of time. So, the sum of time and antitime is always zero.

Men, living in world α, perceive phenomena that occur according to the arrow of time. However, humans can also perceive some objects of the γ world which can equally be interpreted as moving according to the inverted arrow of time, or as moving as superluminal phenomena when the direction of the arrow of time is taken as the standard property.

According to a rigorous mathematical formulation, the two options are formally equivalent, although - for symmetry reasons - the principle of conservation of the flow of time-antitime seems more "beautiful".

Consider, however, Fig. 4.1 (i.e., Fig. 6.5 i.e. 9.1). We predicted that the materialization of a photon depends on an endemic, although unobserved, property of the β-world. Likewise, we introduced the "exotic" hypothesis that, due to an endemic property - both in the α world and in the γ world - every emp or antiemp sample constantly releases in every direction a "rubber band" - or a "double helix" - composed of miniphotons of opposite polarity, i.e., miniphotons and antiminiphotons. These emissions remain "*virtual*" unless they are captured by some other sample of emp or antiemp.

Furthermore, emp samples in the α world only perceive miniphotons, while antiemp samples in the γ world only perceive anti-miniphotons. Therefore, what is predicted by the previous argument confirms the speculative hypothesis of the "rubber band" - or "double helix".

Once again, we can conclude that antimatter can be equivalently considered as a world with an inverted arrow of time or as a superluminal world, consistently with Fig. 22.1f. *A priori*, there seems to be no reason to suggest that one of the two mnemonic models should be preferred to the other.

However, do superluminal phenomena exist or not? A possible answer can be linked to cosmological elements, as we will show in Chapter 23.

Chapter 23

Superluminal Phenomena
Topology of the Worlds α, β and γ
The interaction between Different Objects

"The unification of humanity is achieved through utmost inner freedom, i.e., through a full conscience, a real profound knowledge, a feeling of fraternity, communion of interests and spiritual growth. 'The world can afford to manifest itself - or rather, we can safely claim, it exists - only through conscience' *Schrödinger (1951). The world gets its full reality only from the emergence of its data within the human consciousness of the researcher… Nature must be considered as a single, undivided whole in perennial transformation. It must be carefully observed with no preconceived feeling derived from abstract speculation."*

[*Terra, pianeti e spirito. Il fenomeno geologico come simbolo di un fondamento immateriale,* (p. 161)]

Forese Carlo Wezel (1936-2018)

[*Academician of Lincei,* Emeritus Professor of Stratigraphy at *Urbino University,* (Italy)]

……………………………………….

The term "topology" has a very wide meaning. According to the *Treccani Encyclopedia*, in mathematics, topology is a sector of geometry, which can be defined as the study of those properties of figures which do not vary by subjecting the figures themselves to continuous deformations (which do not cause breakages or superposition of points). Therefore, no notion of measurement (lengths, areas, amplitudes, etc.) is part of topology. For instance, a circle is, from a topological point of view, equivalent to a square, but not to an annulus - or a ball is topologically the same as a cube, but is not topologically the same as a life-buoy swimmers, etc.

The topology of the three worlds α, β and γ causes some perplexity, i.e., we wonder how the three worlds can possibly intertwine with one another.

The world β is an omnipresent feature, a truly absolute reference system. Every limited space-time domain, such as, e.g., a monad, or a galaxy (like our galaxy), etc. can be conceived as embedded mainly in the α world, or symmetrically in the γ world concerning antigalaxies.

We ask, however, whether a galaxy composed of antimatter (i.e., an antigalaxy) can emit photons that we can observe - because it is reasonable to ask whether antigalaxies emit antiminiphotons that we cannot perceive. On the other hand, according to the "rubber band" – or "double helix" – argument, antigalaxies release miniphotons of both positive and negative parity. Hence, we should detect their miniphotons.

Our doubt also concerns the gravitational interaction, which could exist - or not - between galaxies and antigalaxies. In any case, for brevity's sake, every domain - where galaxies exist - or every domain - where antigalaxies exist - i.e., both in the α world and in the γ world, can be defined as a "*bubble*" of the α world or of the γ world, respectively.

Consider Alice (or A), Bob (or B), Charlie (or C), Deborah (or D) and Frank (or F), where A, B and C live in the world α, D lives in the world β and F lives in the γ world. Remember the argument of Fig. 4.1 (or Fig. 6.5, i.e., Fig. 9.1), which represents matter and antimatter as "exhalations" which subsequently eventually return to the β world.

The phenomenon occurs through a repeated process, due to an endemic property of the β world unknown to us. The timing of the process can ultimately be measured by A, B, C or F, while it cannot be detected by D, as D "rides a photon", so she cannot measure time or antitime.

The concern is rather about whether A, B or C, which are in a given bubble of world α, can possibly detect - or not - phenomena that occur in a different bubble of world α.

In other words, we live in our galaxy. Can we observe phenomena inside another galaxy that partake to a different bubble in world α? That is, we can imagine a hierarchy of bubbles of increasing order, depending on the number of crossings of the "*brane*" (see Chapter 22) of the β worlds that are necessary for a transit from a given bubble of world α to another bubble of world α, etc.

Bubbles of different orders can appear, perhaps, as objects moving at relative speeds exceeding several multiples of the speed of light c. In fact, we can hypothesize – and tentatively imagine – that, in that case, Fig. 22.3 should look like Fig. 23.1. That is, galaxies that – on the basis of the current interpretation based on the (erroneous) standard relativistic formula for the Doppler effect – are believed to expand at a given speed $\mathbf{v}$, should actually implode at a speed $2c + |\mathbf{v}|$, although they partake to a bubble of world α which is of hierarchical order increased by one. The same argument applies to hierarchies increased by several units, etc.

However, we should emphasize the symbolic and perhaps unrealistic nature of Figs. 22.3 and 23.1. In fact, the abscissa axis is a mixture of alternating segments representing time and antitime (as in Fig. 23.1). That is, when galaxies and antigalaxies annihilate and enter the β "*brane*" they end up materializing again as new galaxies and antigalaxies that enter the α world and the γ world again, etc.

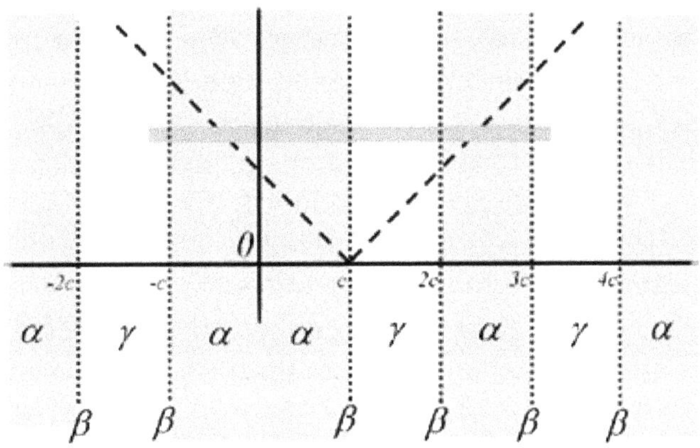

Fig. 23.1.The hierarchy of bubbles of alternately α-worlds and γ-worlds can be, conventionally and symbolically, unified into a common scheme by considering a sequence of temporal and antitemporal axes. In every pair, the temporal-antitemporal axes "touch" each other at the same point (not measurable, but existing) located in their common 3D "*brane*" of the β world. Every conventional composite light ray - which hypothetically moves through such a sequence of bubbles of the α-world and the γ-world alternately with a time arrow of opposite direction - is a reasonable candidate for being the image of our actual physical perception of galaxies and antigalaxies. Differently stated, in principle, the photons we receive from distant space can be released from both galaxies and antigalaxies, having a relative velocity $|\mathbf{v}|$, measured in our reference system which is embedded in our world α. But the primary source of the photon can also be imagined as superluminal. Therefore, the observed redshift, according to the logic of Fig. 11.2, does not necessarily imply an exploding Universe. Rather, there may exist some galaxies or antigalaxies that implode at some superluminal velocity and which, by the current standard, we interpret as celestial objects derived from an original big-bang. This basic epistemological dilemma, however, is unsolvable, since both hypotheses are equally self-consistent, even if one or the other hypothesis has different implications for cosmological models. See text.

Strictly speaking, it is conceptually impossible - and it is disquieting to us - for any observer to distinguish between the two possibilities, since an observer is incapable of recognizing the hierarchy of any galaxy. Therefore, choosing one of the two possibilities is only a matter of principle. It only expresses an indeterminacy, as the empirical constraint prohibits freeing us from this uncertainty.

Maybe this hierarchy doesn't even exist. However, in principle, it can exist and - in fairness - we cannot simply reject the existence of an undesirable conclusion.

In summary, the hypothesis that the speed of light c is the maximum possible asymptotic speed was Einstein's astute idea. As long as we remain in our α-world, no object can be launched at light speed by any finite energy source. However, this assumption applies inside world α. On the other hand, in principle, it is possible that superluminal phenomena also exist, although we can perceive their existence only indirectly through a hierarchy of "bubbles" of α worlds, interspersed with "bubbles" of γ worlds, separated by "*branes*" of worlds β.

On the other hand, consider galaxies or antigalaxies imploding at superluminal velocity. It is reasonable to suppose that they will never enter our world α. In fact, they should cross one or more "*branes*". Every time a galaxy or an antigalaxy crosses a "*brane*", it enters a world with the opposite arrow of time. That is, when a galaxy or antigalaxy crosses a "*brane,*" it reverses its motion. Therefore, it will never enter our world α which appears to be "protected" by the shield represented by one or several "*branes*".

In other words, the process of figures 22.3 and 23.1 seems almost an "oscillatory" phenomenon between worlds with opposite arrow of time.

A related question is the location of dark matter and dark energy within the Universe. It is now generally stated that the existence of the dark emp manifests itself with the apparently increasing expansion speed of galaxies. The estimate of the velocity of galaxies is based, however, on the incorrect relativistic formula for the Doppler effect. On the other hand, no one can say whether the correct (currently unknown) formula for the Doppler effect ultimately gives a different result from current estimates.

Furthermore, the existence of the dark emp itself seems a very reasonable hypothesis, considering the strong limitation we suffer due to the empirical constraint. The β world is ubiquitous, and it is tempting to conclude that everything that exists in the β world can be considered to be composed primarily of dark emp.

In any case, if we assume that emp is positive in the world α and negative in the world γ, emp in the world β behaves as something which is not definable - just as no value can be defined (Fig. 22.4) for the function represented at the point corresponding to the abscissa of the asymptote, i.e., to the "*brane*". This amounts to claiming that a mysterious and unknown property, endemic to the β worlds, is responsible for the "exhalation" of the α world and the γ world.

It seems reasonable - or at least plausible - to suppose that, owing to the conservation of emp in every interaction, the content of emp in the world β is changed only when something "annihilates", i.e., because of something that previously existed in the world α or in world γ. In other words, the emp simply disappears from the world α and symmetrically antiemp from the world γ, as they "annihilate" and return into world β.

On the other hand, we are completely incapable of knowing what laws govern world β. We only know that it has an endemic property that materializes "*non-degraded*" emp into "*degraded*" emp. According to A, B, C, and/or F, whenever an annihilation process occurs, some amount of positive and negative emp apparently cancels and compensates with each other. So, the final emp input into world β is formally zero, although it is just emp and antiemp transformed from "*degraded*" emp to "*non-degraded*" emp.

Furthermore, concerning dark emp, since we are "blind" to different manifestations of emp within world α, and mainly within world γ, we have no difficulty in speculating about the existence of all dark emp of which possibly we need, which could also be located directly within the α world, and mainly within the γ world, without needing to appeal to some exotic hypothesis concerning the β world. Ultimately, we are prohibited from knowing how much emp exists in the Universe.

This conclusion originates, however, a new question about the nature of the miniphoton, namely, whether a photon should be considered as something participating in the β-world - in which case the asymptotic speed c should be identified with the speed of light. In fact, an alternative possibility is that a photon is something moving at *almost* that asymptotic speed, although slightly less than c, thus partaking in the world α. But, in the case in which the photon does not move at the assumed asymptotic speed c, we ask ourselves if other particles exist. If they exist, we ask ourselves whether they move at this asymptotic speed c, or at a slightly lower speed.

For simplicity, in the following, we will therefore call "*elementary granules*" those elementary mini-grains that are the typical hypothesized inhabitants of the β world, whether they are miniphotons or other.

The interaction among different objects. An axiomatic general mechanism

Two given "objects" represent two entities that "exist", i.e., the two objects have a given emp, which can be either "*non-degraded*", or even "*degraded*". A formal and general axiomatic hypothesis might be the following. *SEA*

(*Steady Exhalation and Annihilation*) was defined in Chapter 14, assuming that every existing object releases, through *SEA*, virtual miniphotons of both parities. In Chapter 14 this mechanism was called "*rubber band*", although on the basis of an "exotic" hypothesis that seemed reasonable. The "*rubber band*" was then reinterpreted in terms of a "*double helix*" (see Fig. 10.4).

We now want to attempt to speculate also from a more general viewpoint. In any case, objects endemically release virtual miniphotons of both parities (like the "*rubber band*" of Fig. 6.2 or 14.1). We then hypothesized that the receiver plays a critical role, as it discriminates whether virtual miniphotons "materialize" that are of one parity or the other. Thus, the α-world receivers should detect only miniphotons and the γ-world receivers only anti-miniphotons.

In contrast, as far as our discussion is concerned, it seems more "beautiful" to substantially modify the *SEA* hypothesis into a simpler form. In particular, in certain respects, the concept of "*virtual*" objects - which seems "exotic" - can be traced back to a simple direct equilibrium of both emp and generalized momentum. In fact, the balance of the quantity of emp occurs through a statistical compensation between "elementary granules" which are, altogether, both released and captured.

The reader may judge the *SEA* concept to be revolutionary, while in reality it has a realistic physical content. In fact - as a clarifying example - let's remember the case of the gravitational interaction between the Sun and a planet. The gravitational field of the Sun manifests itself only if a planet is in the right place and at the right time, so that the gravitational force can be effective on it. When such an interaction occurs, it is stated that a certain potential energy originating from the Sun is transformed into kinetic energy of the planet.

However, exactly the same phenomenon can be described by different words as follows, and avoiding the concept of force and field. The Sun releases "*elementary granules*" (for the moment forget whether or not they have parity), which must be considered as "*virtual*" entities that can possibly be captured or materialized by a receiver. If a suitable receiver exists, an effective transfer of energy and generalized momentum occurs at the expense of the action originating from the Sun.

That is, the Sun generates potential energy - or releases "*elementary granules*" - which is a different way of saying the same thing. On the other hand and according to one of the two definitions, none of these entities can be detected or measured, unless an object is suitably located in spacetime, so that it can perceive the effect of the presence of a "*virtual*" entity that otherwise is undetectable.

That is, you need a "receiver" that "captures" (or "materializes") the solar signal. In this way, the "receiver" becomes the "detector" of the solar signal. Therefore, the *SEA* hypothesis is not conceptually revolutionary.

Compared to the simple example dealing with solar gravitation, the novelty contained in *SEA* is rather that the *"elementary granules"* of any type have a parity, which is the necessary consequence of the intrinsic complex conjugated nature of reality and - in fact, this is already implicit in d'Alembert's equation (see Chapter 6) etc. Whenever *"elementary granules"* are captured by a receiver, they transform emp into kinetic energy and - according to Newton's dynamics - these *"elementary granules"* are called potential energy.

However, from the viewpoint of our logical journey, such a constant release implied by *SEA* seems "exotic" because it speculates on an immaterial and unobservable (i.e., *"virtual"*) quantity. This is the meaning of the *"rubber band"* – or *"double helix"* – with miniphotons and anti-miniphotons which, however, are only perceived by receivers in the α world or in the γ world, respectively. Otherwise, these miniphotons of both parities are ignored, as they are *"virtual"*.

In other words, the "exotic" hypothesis does not so much concern the hypothesis of the emission of *"elementary granules"*, but rather the existence of specular worlds, i.e., the existence of twin emp of matter and antimatter, which ultimately annihilate each other and they enter the world of photons, i.e., the β-world, from which they can eventually materialize again, etc.

This "exotic" hypothesis is, however, a rigorous necessity required by the observations of matter and antimatter. Furthermore, this axiomatics allows getting rid of several paradoxes of theoretical physics, despite the fact that such a concept may seem *ad hoc*. Therefore, one can perhaps hypothesize a better mechanism than *SEA*. But, in any case, we need to justify a process such as the one shown in Fig. 4.1 (i.e. Fig. 6.5 or 9.1), even when the materialized pair of objects is not $e^- - e^+$ (or any other particle-antiparticle pair). In any case, it is necessary to hypothesize a pair of miniphotons, or in general of *"elementary granules"*, of opposite parity.

Indeed, all of this is very reminiscent of the aforementioned concept of endemic release of miniphotons and anti-miniphotons to make up the *"rubber band"* or *"double helix"*. Even further, the *SEA* mechanism refers to miniphotons - and also, possibly, to other *"elementary granules"*, including the particle/antiparticle pair, $e^- - e^+$, etc. That is, the *"rubber band"* – or *"double helix"* – is not just composed of miniphotons. In fact, *SEA* is an endemic process that refers to every type of *"elementary granules"*, even (if they exist) other than miniphotons.

For instance, the process is reminiscent of the observed case of the so-called meson oscillation $B_d^0 - \overline{B_d^0}$ which implies a process that repeats itself at a rate of $\sim 19\, psec^{-1}$, where *psec* denotes a "*pico-second*" or a thousandth of a billionth of a second. This oscillation was measured - and published in 2006 by Abazov and others - who showed how the particle, known as the B_d^0 meson, was observed materializing into a particle-antiparticle pair. Thus, the particle-antiparticle couple annihilates; then materializes, again, etc. The process is cyclic and very rapid. The repetition rate has been measured at $\sim 19\, psec^{-1}$, i.e., the process repeats $\sim 50\ billion$ times in one second.

However, concerning *SEA* and the "*elementary granules*" of opposite parity, to our knowledge, no argument seems able to provide any indication on the speed of the "*elementary granules*". That is, the phenomenon is constant and endemic, unpredictable, identical to the mysterious endemic property whereby some emp in the β-world materializes in matter and antimatter. Indeed, materialization from the world β is also a corollary of *SEA*. On the other hand, due to the principle that nothing can be infinite (see Chapter 2), a finite speed must exist both in the space and time of world α and in the antitime of the world γ (there is no problem in the β world where time and antitime cannot be defined).

In other words, the *SEA* hypothesis is equivalent to stating that permanently and everywhere throughout the entire spacetime, every entity that exists – i.e., every entity that has a non-zero emp content - is always the manifestation of some sort of "exhalation" of emp from the β world, which also implies the exhalation of the antiemp into the complex-conjugate world (see Chapter 19).

It should be stressed that "spacetime" is an abstract concept, a mathematical algorithm - a logical "shelf" imagined by man (see Chapter 3) by means of mere abstraction - which therefore by definition per se also concerns "infinite" (or, rather, unlimited) intervals of space and time, while the existence of emp implies only finite quantities. The β world is an omnipresent absolute reference system that exists concretely everywhere, except, perhaps, in the "*vacuum*".

In fact, we cannot know whether "*vacuum*" is a mere paradoxical abstraction or rather a real characteristic. The existence or otherwise of "*vacuum*" depends on whether the property of materialization of emp from world β is omnipresent - and in this case within which (limited?) extension of spacetime - or, rather, does materialization of emp occur only in a selected domain in the β world?

But it is not possible to identify a selected site in a timeless fashion. That is, there is no limitation in space and time in the world β. In other words, it makes no sense in itself to ask the question whether the world β has a finite or "infinite" extension both in space and in time or antitime. The meaning of "finite" or "infinite" space makes sense only in the world α or in the world γ, while it has no meaning in the world β. It makes no sense to ask the question whether the "absolute" Universe is finite or infinite

In any case, the process of "exhalation" removes, from the world β, one "*elementary granule*" of one parity at a time, together with an "*elementary granule*" of the opposite parity. Let's say that in this way the process "extracts" a pair of specular "*elementary granules*", or twins, of opposite parity, which must be inserted into two conjugated complex spaces.

The total balance – whether of emp, or of generalized momentum $\vec{P}$ or of generalized antimomentum $\vec{\vec{\boldsymbol{P}}}$ – is involved in the process of "exhalation" (we used a non-bold, or bold, mathematical symbol for two objects which refer to world α and world γ, respectively). This requirement is perfectly met by the exactly opposite sign of emp of $\vec{P}$ and $\vec{\vec{\boldsymbol{P}}}$. It should be emphasized that such an equilibrium of E as well as of $\vec{P}$ and $\vec{\vec{\boldsymbol{P}}}$ can only be assessed based on observations made by A, B, C and F, which use their respective sign conventions, while - as mentioned above - D it is located inside the β world, or within a "*brane*", and can measure nothing, not even space and time or antitime.

At the same time, when "exhalation" has been activated, the two "*elementary granules*" – at every site and at every instant of time – can eventually recombine and sink, as in a funnel, thus re-entering their parent β-world.

However, recombination and precipitation into the parent β-world can only occur in the case where no other object ultimately captures one of the twin "*elementary granules*". In this case, the captured "*elementary granule*" changes the equilibrium of the capturing object which refers to emp and to $\vec{P}$ and $\vec{\vec{\boldsymbol{P}}}$. As for the one remaining "*elementary granule*" - since it remains unpaired - it moves somewhere in the complex-conjugate space to be finally captured by some other object. The concept is intuitively illustrated in Fig. 23.2.

Two additional basic physical aspects must be pointed out.

A first point concerns the nature of the transporter, i.e., the "*elementary granule*". As a first possibility, this "fundamental" vector is the typical physical ingredient for phenomena that occur in the world β. The "*elementary granule*", when observed within the α or γ world, moves at the asymptotic speed c and for the time being, let us hypothesize that it is represented only by the miniphoton. At a later appropriate time, this hypothesis can perhaps be eventually replaced by assuming another type of "*elementary granule*".

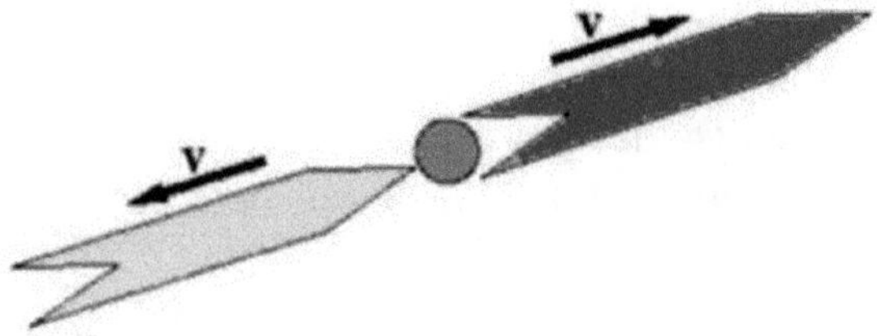

Fig. 23.2. Every existing object (symbolically represented here by a circle) is characterized by an emp content. The *SEA* hypothesis assumes that this object releases in every direction and at every instant of time two twin "*elementary granules*" of opposite parity (in reality the process occurs according to a certain discrete timing and spacing, due to the principle that there is nothing infinite, and also at a speed that we perhaps cannot measure; see below). In the present sketch, the velocity **v** of every elementary granule is referred to the absolute reference system, i.e., to æther, according to the observations made by A, B or C, and is indicated by a solid arrow. The two vectors **v** are therefore directed in the opposite direction. However, the generalized moment $\vec{P}$ and $\vec{\vec{\mathbf{P}}}$, associated with every "*elementary granule*" of one of the two polarities, are simply parallel to **v**, which is positive or negative, respectively, in the α and γ worlds. The different sign of $\vec{P}$ or $\vec{\vec{\mathbf{P}}}$ is shown by the directions of the large arrows. This means that only one "*elementary granule*" (here shown as upper arrow) has the same direction as **v**, which is the same direction as $\vec{P}$ (as it occurs in Newtonian mechanics), while the other "*elementary granule*" (here shown as lower arrow) has a $\vec{\vec{\mathbf{P}}}$ oriented away from **v**, where **v** is observed and measured from A. See text.

This "*elementary granule*" requires a subtle additional precision. One wonders whether there is any need, required by the observations, to introduce an additional carrier, different from miniphotons or from the "*elementary granule*" of a single type. Or should we instead hypothesize a second type of "*elementary granule*" (e.g., for gravitation)? According to the current knowledge, it does not appear that observations require such an additional hypothesis.

The second point concerns a new physical hypothesis, which has profound consequences, and can be called "*electrical hypothesis*". That is, it is assumed that a fundamental quantum number of every particle or antiparticle is the intrinsic ability to capture only miniphotons that are "incomplete", i.e., that are "*lame*", and that both lack identical pairs of "*elements*". Furthermore, this different capacity must be identified with the positive or negative electric charge of the particle or antiparticle (see Chapter 13).

In other words, every given particle (or antiparticle) has a given electrical quantum number, while its respective antiparticle (particle) has the respective opposite electrical quantum number. There is, however, no requirement that a given particle must have one electrical "*virtue*" or another. That is, a given miniphoton parity is not necessarily associated with either the α-world or the γ-world, just as positive or negative electric charge is typical neither of particles nor of antiparticles alone. Indeed, both positive and negative particles and/or antiparticles are observed. That is, matter and antimatter on one hand, and miniphotons of different parity on the other, are mutually independent events that are typical of natural reality. They represent the complex-conjugated nature of reality.

Furthermore, consider the double nature of the interaction of miniphotons, according to the axioms concerning *(i)* the electric charge, and *(ii)* the magnetic field of an electric current, and electromagnetic induction (see Chapter 14 and Chapter 15 and *passim* elsewhere). Furthermore, consider that a particle is a monad, and - due to the empirical constraint - every monad can be considered as being composed of monads of smaller dimension (see Chapter 4).

Therefore, in particular, a neutral particle can be tentatively considered as a composition of at least two monads of smaller dimension, each with a different electrical "*virtue*". For instance, the atomic nucleus is made up of protons and neutrons, etc. That is, if this hypothesis is correct, every neutral particle or neutral antiparticle should always be considered as the combination of two smaller components, i.e., of more elementary particles or antiparticles, one of them with positive electric charge and the other with negative electric charge.

Indeed, if we want to allow for any possible interaction of neutral particles - which is different from the aforementioned interaction between smaller charged monads - this possible additional occurrence requires a further *ad hoc* hypothesis. This violates Ockham's razor, however (see Chapter 14).

Therefore, unless a convincing requirement is found in the observations, the working hypothesis that seems "simplest" must be considered. It seems reasonable to ask whether the resulting interaction of the two-component monad - with a total neutral electric charge - should be identified with the current so-called "gravitational" interaction, as hypothesized in the axiom of gravitation formulated in Chapter 15. Hence, only a subtle doubt remains. However, the unexpected conclusion is that, if the simplest hypothesis is correct, gravitation would be just a "thermodynamic" expression of electromagnetic interaction. A fuller discussion is provided in Chapter 25.

Chapter 24

Superluminal Phenomena
Astrophysical and Cosmological Implications

"Don't be concerned: dream great! … Do not lose courage and aim high!"

Pope Francis [in 2014, addressing the young ones]

……………………………………….

Suppose that, after applying a correct formula for the Doppler effect, two currently generally reported facts are confirmed, namely *(i)* the expansion of the Universe and *(ii)* relatively farther away objects have a higher escape velocity (or *red shift*).

This belief is based on the fact that we guess that, when we observe the *red shift*, we are correct if we interpret the evidence in terms of a light source moving away from us. However, according to Figs. 22.3 and 23.1, the same observational evidence can be equivalently interpreted in terms of celestial objects imploding towards us at a speed $|\mathbf{v}| > 2c$.

That is, the direction of $\mathbf{v}$ is reversed: perhaps is this equivalent to a reversal of the arrow of time? (see below). That is, the expansion of the Universe, the Big-bang, etc., they would have to be converted into a huge imploding Universe. On the other hand, such a big-crunch (a term denoting the opposite of big-bang) in a superluminal world must cross through one or a few β-*branes*, where the implosion becomes an explosion (and viceversa). In other words, galaxies and antigalaxies annihilate themselves and enter the β world, to later rematerialize and generate new galaxies and antigalaxies. That is, the galaxies must transform in a cyclical, almost oscillatory, way.

Furthermore, it is argued that the evidence for the dark emp is based on the fact that the expanding Universe seems to imply that the rate of expansion is paradoxically increasing, rather than - as would seem intuitively more logical - decreasing. Dark energy should then justify and solve this paradox.

This means that in Figs. 22.3 and 23.1 the redshift increases, or the points on the graph move upwards. Conversely, if we interpret this evidence in terms of an imploding Universe with $|\mathbf{v}| > 2c$, we conclude that the implosion speed increases with respect to time, consistent with what one should intuitively expect due to the simple effect of the observed emp.

That is, the paradox no longer seems to exist concerning the existence of dark matter and dark energy. On the other hand, dark emp should exist, since we can know almost exclusively only matter that emits photons – i.e., we can only observe matter that is reasonably cold to allow the existence of electrons captured in atomic orbitals.

If the Universe is composed primarily of hot matter, and if the largest percentage of matter is composed of "naked" nuclei (see Chapter 1), we cannot see the corresponding large amount of emp. Consider that, because of the implication of Fig. 21.3, we have an equal ability to observe celestial objects imploding by emission or mirror Doppler effect.

Instead, suppose we are dealing with several cases involving celestial objects moving away from us at a speed $|\mathbf{v}| > c$. In fact, in the case of the emission Doppler effect, we should recognize the red-shifts with $\lambda'/\lambda \geq 2$. However, at the moment, this circumstance seems difficult to finally evaluate in our observations (perhaps, we should look for it critically and evaluate the consistency of this possibility with available observations).

By contrast, the mirror Doppler effect is obscured from us. In both cases, we conclude that, if we assume that the Universe is full of emp moving at $|\mathbf{v}| > 2c$, it appears that - owing to this dimming - compared to exploding celestial objects, we are biased in detecting galaxies imploding. In any case, we wonder whether the mirror Doppler effect actually exists.

In any case, concerning the standard cosmological models, consider that "time" is not absolute in the Universe. Therefore, two complex symmetric conjugated universes exist, with opposite temporal arrows, of opposite parity, composed of "*degraded*" emp, i.e., of matter and antimatter, respectively. World α is described by "time" and world γ by "antitime". Furthermore, the β world is "*non-degraded*" emp which looks like electromagnetic radiation in both α world and γ world.

Note, however, that matter and antimatter, i.e. α-world and γ-world, together with β-world, coexist everywhere and are closely interconnected with one another through the *SEA* process (see Chapter 23). Humans are made of matter and live within – and detect phenomena that occur within – the α world. Perhaps there is no need even to hypothesize that antigalaxies composed of antimatter also exist, since matter and antimatter are, in fact, omnipresent.

In summary, if we assume that everything must propagate with velocity $|\mathbf{v}| \leq c$, we get the currently generally accepted and agreed-upon model of an exploding Universe, of big-bang cosmology and of the need for a large amount of dark matter to justify the increase temporal of the expansion speed, etc. This cosmological model thus implies a situation that seems paradoxical. For example, what is the initial instant of time of the big-bang? How can such an assumption fit into the non-absolute nature of "time"? What is the physical meaning of time's arrow? What had come before? Etc.

In contrast, if we assume that some emp in the Universe can move at $|\mathbf{v}| > c$, we must conclude that we are biased observers, because we can observe mainly exploding emp. Thus, we are likely to lose evidence of a relevant fraction of imploding emp. This fraction depends (as shown in Fig. 21.3) on the unknown fraction - in our Universe - of the mirror Doppler effect as opposed to the emission Doppler effect. However, we ask ourselves whether the mirror Doppler effect makes sense.

On the other hand, a large amount of what we interpret in terms of an exploding Universe can be interpreted in terms of an imploding Universe with a suitable $|\mathbf{v}| > 2c$. Only Ockham's razor (see Chapter 14) can help choose the apparently "simplest" model, although, ultimately, the empirical constraint prohibits us from deducing any final choice between the two possibilities.

The same concern can also be clarified as follows. If, as a matter of principle, we can observe nothing that ultimately moves at a $|\mathbf{v}| > c$, we can observe no galaxy moving at a velocity $|\mathbf{v}| > 2c$, and Fig. 21.3 raises no ambiguity.

However, we ask ourselves whether we are sure that the redshift of a galaxy is not just an exception that shows how we also observe objects moving at $|\mathbf{v}| > c$? It is not possible to answer.

In any case, we need to know the correct law for the Doppler effect first. In fact, the concern over the Doppler effect is even more disturbing. In principle, both types of Doppler effects, mirror and emission, can be observed, measured and checked in the laboratory - provided we know how to obtain a photon source or a mirror that moves at a $|\mathbf{v}|$ comparable to $|c|$.

Concerning the ambient Doppler effect, we have no adequate information. When it comes to astrophysical applications, we should take into account every possible type of currently unknown environmental effect, which acts both on the light source and also on the photons as they cross through intergalactic or interstellar space.

As far as laboratory measurements are concerned, it is difficult to generate, e.g., a suitable gravitational force. It can be assumed, e.g., the chronology of cases according to which we eventually detect gravitational waves by means either of a mechanical Weber antenna or of a *LIGO* or *VIRGO* experiment (i.e., Michelson-Morley type; see Chapter 17), etc. and/or, maybe, of the natural antenna of some structure (e.g., perhaps, through a mountain range, etc.).

Instead, suppose that, simultaneously, we fail to detect any gravitational wave by means of any "transient Michelson-Morley experiment." In this way, we could verify that both the length of the arms of the interferometer, and the atoms of the laser source, react identically to the incoming gravitational wave.

This would be one way to test an atom's response to the impact of a gravitational signal.

In contrast, several cases are now available of observations of massive gravitational perturbations monitored by *LIGO/VIRGO* experimental devices. So, compared to atom-monads, large-scale structures respond differently - like the arms of a huge Michelson-Morley interferometer.

In any case, concerning astrophysical studies, all analyses carried out in the past should be reanalyzed, avoiding any reference to the relativistic formula for the Doppler effect. Note that both (21.25) and (21.26) also take into account a certain amount of the ambient Doppler effect.

Note, however, that the general perspective is frustrating, since we have to use a Doppler effect that is heavily distorted by the empirical constraint and by some unknown physical law. In fact, galaxies and antigalaxies (perhaps, see anyway below) should interact gravitationally because, due to the *SEA*, they both release "*elementary granules*" of opposite polarity: galaxies only capture "*elementary granules*" of positive polarity, and antigalaxies only capture "*elementary granules*" of negative polarity.

The evidence of dark emp also requires critical rethinking. A previous hypothesis was that dark emp should be compared (at least in part) to neutral particles, i.e., to non-ionizing radiation, which therefore cannot be directly detected by humans.

However, the concept of dark emp should now, perhaps, be expanded to include all emps that move through the Universe at a certain speed $|\mathbf{v}| > c$. On the other hand, does it make sense to talk about something that we can't detect? To the extent that we have no evidence (even indirect) - and to the extent that we are dealing with something we cannot measure - it makes, perhaps, as much sense as talking about "*vacuum*", or a "spiritual receiver" etc.

In any case, this is an inevitable drawback of the entire cognitive process of humans when faced with Nature. Ultimately, men are much more "blind" than we have very presumptuously believed throughout past history.

Even the current general belief, by current standards of "excellent" science, is based on the assumption of an omnipotent power of "science". This sentiment is still a child of the trend started by our brilliant (and egocentric) ancestors and thinkers of ancient Greece. The ancient Greeks believed that humans knew the perfect laws of logic, the same laws that "*were applied by God during the creation of the world.*"

On the contrary, we must realize that we cannot discover - and we cannot know - the true ultimate "fundamental" laws of Nature. We can only exploit partial knowledge based on the observational evidence that - with respect to every specific monad that we want to investigate - we are objectively allowed to collect, depending on the size of the human body within spacetime.

This last conclusion, however, is certainly heretical for the current canons of theoretical physics. But it is not as negative as positivists claim. Indeed, imagination and human creativity can play an important role in our knowledge. In a certain sense, we can work around the limitations of the empirical constraint by means of our imagination and creativity, and we can imagine some hidden parameters. On the other hand, our speculation must be subsequently compared with the observation of the "*virtues*" of every monad.

In any case, our formulation of the hidden parameters - even when it is confirmed by observations - is based on an arbitrary conceptual scheme, i.e., on a "simple" or "beautiful" mnemonic model. On the one hand, such a model can only be considered a realistic - and possible - working hypothesis when it is supported by observational tests. On the other hand, such a formulation of the hidden parameters is subjective and cannot be unique. Indeed, someone can always imagine a different, "simple" and "beautiful" model that fits the observations, even if in the end it is very different from the previous model.

Note that, in this sense, we can remember, e.g., ancient Greek mythology which can be compared to a set of easy-to-remember astronomical rules, capable of managing a complicated geodesic network made up of temples and sanctuaries, etc. That is, at that time, mythology could be compared to a real and very effective symbolic representation made up of hidden parameters. The stories of mythological characters were excellent mnemonic devices, suitable for a society that generally suffered from widespread illiteracy.

Different theoretical schemes or models (such as, perhaps, "string" theories, "*brane*" theories, etc.) based on a multidimensional space, etc., can be considered - in a certain sense - as wonderful exercises made by human imagination. However, no model is intrinsically better than another. It can be "simpler" or more "beautiful". Certainly, Ockham's razor can help avoid any overly complicated theories.

On the other hand, generally, no product of human creativity is better than another. Every product is a wonderful creation of the intellect, just as different theoretical frameworks may be suitable to frame our knowledge in a "simple" and "beautiful" scheme or framework that should be easy to remember or to use to predict the behavior of natural phenomena.

In any case, one should never forget that Nature per se is much more complicated than any "simple" and "beautiful" scheme that we can conceive.

Chapter 25

Superluminal Phenomena
A Summary of "Simple" Hypotheses

"*...your sons and your daughters shall prophesy, and your young men shall see visions, and your old men shall dream dreams ...*"

After the St. Peter speech of Pentecost [Acts 2, 17].

This St. Peter's sentence, which was inspired by the prophet Joel (4th century BC), means that sons and daughters, owing to their fluency and indomitable idealism, must criticize and envisage and look for better living ways. Young men must be enlightened while taking their operative decisions. Old men - who are free from their former duties and responsibilities - owing to their expertise and skill deriving from their long life, must propose "dreams", i.e., new targets or perspectives. This concept of the three ages of man always fascinated great artists. For instance, remind about the enchanting painting by Giorgione (Giorgio Barbarelli da Castelfranco; c.1477/8-1510), known as "*The Three Ages of Man*" or "*Reading a Song*", at present on display at the *Galleria Palatina* in *Palazzo Pitti* in Florence.

....

State-of-the-art

Let's summarize all the previous arguments, and summarize everything as much as possible.

Believe it or not, our working hypothesis seems to be "simple" and attractive. The absolute 3D space is crossed by "*double helices*" - or photons - in the timeless β space. However, we cannot detect them, nor do we know their density in β-space, where, due to a lack of time, we cannot even count them.

It should be emphasized, however, that the concept of space – as a "shelf" suitable for organizing and storing observational knowledge (see Chapter 3), is valid only in the world α and in the world γ. In fact, in the atemporal β world, it is not possible to localize different points in a 3D space, since locating different points would be equivalent to identifying different events, even if it would be impossible to associate different times with different events. Moreover, an event that enters world β from world α or world γ could maintain a recognition label in time or antitime. But this contradicts the strictly atemporal nature of the world β.

This aspect has very important implications. In fact, it is reasonable to ask ourselves the problem of the extension of 3D space in the world α or in the world γ, i.e., to ask ourselves whether a 3D space is "finite" or "infinite". On the contrary, it makes absolutely no sense to ask whether a space associated with the atemporal world β is "finite" or "infinite". It is just not possible to define an extension of the world β.

This aspect also has cosmological implications relating to the big-bang and the big-crunch. In fact, the big-bang model, characterized by a huge explosion from a central point, makes no sense, just as the big-crunch model, characterized by a huge implosion at a single point. Big-bang events and big-crunch events are only measurable in a 3D space in the α-world or the γ-world, but not in the β-world. Therefore, the galaxies and antigalaxies that we observe with our telescopes are generated at sites that, for observers in the α-world or the γ-world, are randomly located in 3D space. Cosmological models must therefore be fundamentally rethought.

When a "*double helix*" splits into two unpaired single "*helices*" of opposite parity, a "materialization" process occurs. Let's say that a previous photon originated a particle and an antiparticle. In general, every particle and antiparticle is associated with a single "*helix*" which is not "*complete*", i.e., it is "*lame*", i.e., the particle has an electric charge and the antiparticle, being associated with the twin "*helix*", has the opposite electric charge.

Both the particle and the antiparticle can possibly experience forward Compton scattering with other photons. The ultimate result is that we observe a particle and an antiparticle with a given kinetic energy, which is the result of the rest mass, or rest antimass, respectively, of the particle and the antiparticle, plus two "*boosters*" attached to both, representing their respective positive and negative kinetic energies.

Particles (or antiparticles) in the α world (or, respectively, in the γ world) interact with each other according to the laws of dynamics (or of antidynamics), exchanging different amounts of "*boosters*", which are equivalent to the exchange of kinetic energy, and this is what we see. This is the process of thermal interaction, which excites atoms, ions, and molecules that release the photons that we observe.

What photons can we detect? Let us consider a particular class of photons generated by the photon factory represented by a monad-atom, or by a monad-ion, or by a monad-molecule. Our eye is an antenna tuned to this kind of photons. Other photons, if they exist, cannot be detected by our eyes. The same is true for a photon sensor, i.e., a photographic plate or a photomultiplier, or whatever, as every sensor has a limited sensitivity to the type of phenomenon foreseen in its design.

We observe particles, and sometimes even antiparticles, because we live in the α world, and we are composed of matter, and we observe photons of a specific parity, i.e., generated by specific photon factories.

When two particles in α world interact, they undergo an attractive or repulsive action depending on whether they share an opposite or the same electric charge. The symmetric phenomenon occurs in the γ world containing antiparticles.

Note that two interacting particles or antiparticles mutually manifest their respective nature - either in the α world or in the γ world - by means of photons, or by means of "*double helices*" characterized by the same "*lame*" peculiarity of the parent particle or antiparticle.

The "*receiver*" of the message represented by the "*double helix*" is sensitive to both "*helices*" of suitable parity, consistently with the world α or world γ which is considered.

A particle represented by a single "*complete*" "*helix*" is a neutral particle (or antiparticle, depending on the parity of the "*helix*"). It interacts with other particles (or antiparticles) through the same mechanism, based on a message that maintains the same "*complete*" or "*lame*" character as the parent particle. For instance, it is well known that neutrons have a higher scattering cross section for slower neutrons, just as if slower neutrons had a greater chance of "seeing" each other.

When a particle encounters an antiparticle, "annihilation" occurs, as two unpaired "*helices*" of opposite parity make up a "*double helix*", which is a β world photon.

Consider a large ensemble of particles (in the world α) composed either of neutral particles or of an equal number of positively and negatively charged particles: they experience gravitational attraction.

Therefore, every galaxy feels the gravitational pull of every other galaxy. The reason lies in the tenuous difference between the attractive and repulsive action of "*lame*" spirals of different or equal electrical charge, respectively. Owing to the same mechanism, photons do not feel the gravitational attraction of other photons, while gravitational (mutual) attraction exists between photons and galaxies or celestial bodies. The symmetric phenomenon occurs for a large ensemble of antiparticles in the γ world.

Let us consider the case of a galaxy and an antigalaxy - or a massive star or planet and an antimassive star or planet. A galaxy and an antigalaxy are generally located at a large relative distance. Whenever they interact with each other they can only experience "annihilation". That is to say, galaxies and antigalaxies (perhaps) do not experience mutual gravitation.

In fact, galaxies and antigalaxies, owing to the *SEA* mechanism, emit "*messengers*" of both parities, and galaxies and antigalaxies capture only the "helices" of their polarity. Therefore, the gravitational interaction could also occur between galaxies and antigalaxies as long as the same mechanism can apply to the small difference between the attractive and repulsive action of "*lame*" spirals of different or equal electrical charge - and furthermore this mechanism must also apply when it comes to "*lame*" spirals originating from a galaxy or an antigalaxy. The issue is subtle and requires further thought.

We therefore refer to a huge extension of 3D space, which contains a multitude of photons (in β world) plus a very large set of particles and antiparticles, in α world and γ world, respectively. Matter and antimatter are organized in a large number of galaxies and antigalaxies, which experience a respective evolution in time or in antitime.

Occasionally, one or more galaxies - and one or more antigalaxies - interact and "annihilate" each other. This is a big-bang that gives rise to an enormous quantity of photons, which eventually generate an enormous quantity of particles and antiparticles, organized into a large quantity of new galaxies and - symmetrically – of new antigalaxies.

The big-bang phase persists until galaxies and antigalaxies ultimately annihilate during a sort of big-crunch that restarts a new big-bang cycle.

This is the timeless tale of our Universe. Perhaps some author will devise a better scheme. The debate is open.

The internal mechanisms of monads - A common law: the "Golden Rule"

Ultimately, two key limitations impede our cognitive capacity: *(i)* the empirical constraint and *(ii)* monads that can only be observed from the outside.

Every monad has an intrinsic law that determines internal evolution which, however, if observed from the outside, cannot be detected by men. On the other hand, there must exist some laws, according to a basic fundamental logic that, perhaps, we can only speculate in terms of a "mnemonic" and "simple" model.

For instance, we observe the monad-atom from the outside, through the "*virtues*", which are the miniphotons or photons that it expels due to the "slingshot effect" etc. Let us consider a monad-atom or a monad-ion: the standard model of the optical spectrum is the only basis of available experimental observations. We can handle our understanding in terms of a "thermodynamic" approach (see Chapter 18 and *passim* other chapters), strictly by means of complex functions, because real functions cannot take into account the conservation of the time-antitime flux. This "thermodynamic" approach leads to the Schrödinger equation. However, due to empirical constraint, we can only observe the amplitude of a complex function, not its phase.

A fundamental drawback is, however, that every time we refer to a subsystem of the entire Universe – such as when we refer to a single monad, however large or small - we introduce a serious approximation. In fact, this is equivalent to "nullifying" the correct perception of the irreversibility of the Universe as a whole. That is, a treatment - which in itself is physically wrong - necessarily implies reversibility.

This approximation is certainly non-physical and it is required by the need for simplicity of the human mind. This is shown in Chapter 7, and the supporting observation is the generalized fractality and the golden ratio φ. In terms of strict rigor, the Universe should be regarded as a single system, and we should not even refer to a single monad alone by itself.

In any case, note that - always according to a "thermodynamic" viewpoint - Maxwell's laws should be valid within all monads that we can observe from the inside. At least, we reasonably assume that Maxwell's laws are also valid on the galactic or galactic superclusters scale, since we have no reason to doubt this hypothesis.

We presume that *SEA* applies to the entire Universe, including galaxies and antigalaxies. Then we assume that, when a "thermodynamic" treatment is applied, we can hope to obtain the same conclusion as found in the monads we can observe from the inside.

In contrast, we are in principle unable to interpret phenomena that occur within monads that we can only observe from the outside. That is, electromagnetic interaction *cannot* be a fundamental law of nature. In fact, it is well known that it *is not valid* within the monad-atom.

However, the *SEA* must always apply. If we want to understand some details of what happens inside the atom-monad, we must break down the atom-monad into smaller-sized monads. We have to separate electrons and nuclei, and we have to study nuclei as monads that can only be observed from the outside.

That is, Maxwell's laws - which no longer hold within the atom-monad - should not hold within the nucleus-monad, where some laws hold which are also different from the laws which hold within the atom-monad or (perhaps) even within the electron-monad.

Similarly, one can break the nucleus into subnuclear monads, and thereby enter the multifaceted realm of various types of elementary particle monads.

The breakdown of an atomic nucleus monad can occur, however, according to specific conservation laws, expressed by "*virtues*", called quantum numbers (electric charge, spin, magnetic moment, strangeness, flavor, charm, etc.) that are indicative of the laws which govern the interior of the nucleus-monad, and of every elementary particle monad. In general, these laws do not necessarily hold - maybe they hold, maybe not - in the atom-monad or within larger monads.

Similarly, we can break a given particle monad, e.g., a proton monad, or a neutron monad, etc. and we investigate how this monad is broken down into smaller monads, according to what conservation laws, etc. That is, a repetitive process is thus carried out, through which we allow ourselves to progressively observe ever finer details of natural reality, even if we will never be able to afford to reach the final detail, the finest of all.

In fact, the empirical constraint is very stringent, because it is related to the dimensions of the human body. On the other hand, perhaps, we can afford to imagine some common logic or rule that, perhaps, is shared by all - or at least by a large number of - monads. In fact, we know that the correct irreversibility of the Universe as a whole is demonstrated by fractality and/or by the golden ratio – and whenever one of these properties can be proven, we can assume that the other should also be valid (see Chapter 7).

In this same sense, surprising help is provided by the mysterious, although highly efficient, analog computer that is innate in our brain (see Chapter 8), as detailed below.

Since an enormous number of observations, even outside the typical realm of physics, exhibit the mysterious golden ratio and sometimes even fractality, it is reasonable to conclude that all these properties are a common feature of natural reality, a universal law, shared by everything that exists and according to some common laws and principles, although ultimately they are unknown to us, due to the empirical constraint.

All of this evidence is clear, although – as far as we can tell – it is considered mysterious. In any case, we assume that some common law of regularity is shared by all natural systems, because the evolution of the Universe is irreversible (or because the arrow of time always has the same direction), and this law can be unknown, although we assume it definitely exists. In this case, all the apparent mysterious evidence seems clear.

Furthermore, natural systems appear to be composed of sets of minimal subsystems, everyone of which is possibly associated with some peculiar number of the Fibonacci sequence. Different subsets of peculiar numbers that are perhaps associated with such a minimal system, should justify the variety of behavior of analogous natural systems (e.g., of different variants of biologically similar trees characterized by differently shaped leaves, etc.).

In addition, it appears that the sensation of regularity is a prerequisite for the survival instinct. Hence, biological systems are very efficient at recognizing the Golden Rule within their perceived sensory signals. In the case of men, this "instinct" manifests as artistic sensitivity (see Chapter 8). This interesting suggestion derives from the consideration of the exact theory of "harmony" in music, which is a science that dates back to prehistory, although based only on a subjective sensation of "beauty".

Thus, the human brain, while perceiving "beauty", is an effective and very fast analog computer that - with an almost negligible time delay - recognizes regularities in sensory perceptions (visual or auditory). More specifically, every step in the convergent process of the Fibonacci sequence should be associated with an existing objective state of the natural system. The convergence towards φ is therefore represented by a sequence of observed states of the system as it searches for equilibrium.

The system, however, never manages to reach the final equilibrium represented by φ, as φ is irrational. In other words, the system evolves in search of a final equilibrium which, however, it can never achieve.

The process stops every time that, owing to a change in some environmental or boundary conditions, a new equilibrium is defined and a new sequence towards φ is therefore started, although the new sequence occurs along states that are different from the previous sequence.

As far as our logical journey is concerned, the final conclusion is that, perhaps, different monads - which can only be observed from the outside and are known only through their respective "*virtues*" - ultimately exhibit the "Golden Rule". In this case, we can reasonably deduce that some unknown internal law is very likely to satisfy the requirements of "additivity" and "principle of variation". That is, we should assume that the unknown laws that govern the interior of these monads should satisfy both the Golden Rule and fractality.

This can perhaps help to understand some facets of phenomena that occur within the monad. For instance, suppose that a spectrum is eventually discovered to be composed of spectral lines that satisfy the "Golden Rule". In this sense, Prof. Stergios Pellis recently demonstrated a connection between the golden ratio and a universal constant that is found in all atomic spectra.

Namely, e.g., for clarity, let us consider the spectrum of the hydrogen atom and the spectrum of the oxygen atom. They can perhaps be minimal systems. Possibly, the spectrum of the OH radical can be derived according to a currently unknown Golden Rule regularity.

The same argument could be applied, e.g., also within the Mendeleev table of elements.

In summary, when dealing with any kind of monad that can only be observed from the outside, we look for the time series of "*virtues*". Empirically, we look for some regularity that in some way can be related to Fibonacci numbers or other series of equivalent numbers, or which, in any case, can be traced back to φ. If it is possible for us to recognize such a type of regularity - i.e., the Golden Rule - we can be reasonably certain that the unknown internal laws governing the monad share the requirement of "additivity" and the "a variation principle".

In a sense, even the fractal properties displayed by the "*virtues*" of a monad are an indication of some regularity that is maintained within a monad, even if the monad can only be observed from the outside. However, as far as the internal processes of a given monad are concerned, every finer details must inevitably be based on speculation, hidden parameters, etc.

In this case, we have two possibilities: either we are able to divide the monad into smaller monads, or we must try to invent a "thermodynamic" model.

As far as "thermodynamic" calculations are concerned, however, it is essential that the axiomatics adapt in any case to the available observations. That is, a crucial aspect concerns the formulation of the "thermodynamic" description of the phenomena which - differently from what is very often reported in the literature - must be carried out using strictly correct statistics. In particular, the procedures usually used for lotteries or for election polls, etc., must be abandoned and one should instead rely on specific statistics, as indicated in Table 18.1.

For instance, "statistics" in the atomic world must be applied to treat the complex function Ψ with the Schrödinger equation, while, due to the empirical constraint, we can only observe the amplitude $|\Psi|$ and not its phase. When "statistics" are directly applied to $|\Psi|$ the results are misleading. Maxwell's laws, Schrödinger's equation, the so-called second quantization and renormalization, are different facets of the general approach in terms of "thermodynamics" and related Taylor series expansion.

In this respect, note that there is no need for a convergence of a series expansion, as it is generally sought for the second quantization, which is based on the hypothesis of continuity (see Chapter 2). That is, natural reality exploits only a few terms of the expansion (which, as a standard, are usually represented symbolically by Feynman graphs) while higher order terms become progressively less likely to occur.

Therefore, in disagreement with objective reality, a formal series diverges only due to a rigorously mathematical - but not physical - hypothesis which implies the extrapolation with equal physical weight of all the terms of the expansion of a continuous function. In contrast, physics applies only to a *limited* number of terms in the series expansion. Therefore, it is wrong to extrapolate mathematically the *infinite* sequence of terms and assume that the extrapolated expansion is also physically significant. This hypothesis is arbitrary, it is not physical and it is misleading, similarly to the hypothesis of continuity, of infinitesimal and infinite quantities. The classical example of the Max Planck hypothesis to explain the black body spectrum is eloquent (see Chapter 2).

The Mendeleev table is the representation of the intrinsic order of electrons inside a monad-atom, which is observed from the outside. Thus, the Mendeleev table is the basic logic for chemistry, for crystalline bonds, for the state of matter (solid, liquid, gas, plasma). Therefore, as we have already highlighted, we ask whether the "Golden Rule" exists within Mendeleev's table. That is, equivalently, since the spectral lines are explained by the Schrödinger equation, we ask if the "Golden Rule" applies within the Schrödinger equation.

In any case, when all the electrons are stripped from the central nucleus, both chemistry and electron bonds are no longer relevant to the state of matter. Therefore, one must refer to monads-nuclei that can only be observed from the outside. We have to rely on the "*virtues*" or "*quantum numbers*" of the monad-nucleus that include the magnetic moment.

The interaction of nuclear magnetic moments cannot be, however, a part of standard electromagnetism. Rather, it is of concern of the second quantization. In fact, the interaction between electric current loops - according to the standard and well-known rules in texts on electromagnetism - is such that, simply depending on a transformation of the reference system, the magnetic and electric fields can change their respective percentage role. This property, however, does not apply to nuclear magnetic moments, which must rather be governed by laws that are generally said to be unknown. Indeed, this logical flaw has some significant implications.

That is, the magnetic moment is a quantum number that requires more in-depth reflection. In fact, if we consider an atomic nucleus as an interaction between "*lame*" volleys - which represent every single proton and every single neutron that compose the given atomic nucleus - we have a convincing explanation of the behavior of the nuclear moment of the atomic nucleus itself (see the Chapter 15).

In any case, when the electron shells are removed or almost completely removed, the nuclei are "*naked*" (see Chapter 1), and nuclear magnetic moments play a major role in the interaction between different monad-nuclei. Hence, matter is in a state that can be called "*magpol*", which is an acronym for magnetic polarization. All magnetic moments are tightly packed together - and oriented along one and the same direction within a sort of "fibrous" structure. This is probably, e.g., the state of the inner core of both the Earth and all large planetary bodies.

In this respect, note that, on the contrary, the inner core of the Earth is generally simplistically defined as "solid" only because it is crossed by seismic shear waves. However, it makes no sense to speak of a "solid" state within a very hot environment in which no crystalline bond can exist. In fact, this flies in the face of every notion of solid-state physics. Instead, it makes more sense to say that matter in the inner core of Earth (and other planets) is in a dark matter type of state, because it doesn't release any directly detectable photons. In fact, the empirical constraint is such that we can only observe matter which releases photons. All other emp forms, which are perhaps the largest fraction in the Universe, are invisible, i.e., they are "dark matter".

Conclusion

We began with the "exotic" hypothesis of the possible existence of superluminal phenomena, and it seemed to be a "beautiful" hypothesis. By today's standards, some galaxies are interpreted as moving outward due to the original big-bang. In contrast, if superluminal phenomena exist, parts of the galaxies should implode inwards at a speed greater than twice the speed of light. In other words, humans can speculate and interpret observations of antigalaxies in terms of an outward flow of antimatter, i.e., an implosion of matter.

It appears conceptually impossible to choose between the two options. If superluminal galaxies exist, some type of "oscillatory" phenomenon should characterize the Universe, where galaxies and antigalaxies move back and forth with respect to our galaxy. In this case, it is as if the arrow of time reverses every time a galaxy or antigalaxy passes through a world *β-brane*. However, the process should occur via annihilation and materialization into new galaxies and antigalaxies.

The choice seems to be optional, or simply a matter of "beauty", since this is an intrinsic property of the topology of the worlds α, β and γ.

In any case, due to the inability to detect the "mirror" Doppler effect (if it exists) for imploding galaxies, humans can detect more exploding emp than imploding ones.

However, a crucial element is the essential use of a correct formula for the Doppler effect, which is fundamental to our understanding in different disciplines. Currently, the correct formula is unknown. A proper study of the so-called *Pioneer* anomaly (see Chapter 16) seems to be the only experiment currently available that can help deduce the correct formula.

The symmetric (or mutually complex-conjugate) generation of world α and world γ must be considered as a manifestation of the conservation law of the time-antitime flow. Because of this conservation law, the required symmetry is such that it always generates specular worlds, twins, complexes-conjugates of each other.

A further fundamental aspect is that, in the atemporal β world, it is not possible to talk about "finite" or "infinite" space. The concept of "finite" or "infinite" space applies only within the world α or world γ.

Therefore, a general axiomatic mechanism for the interaction between different objects, called *SEA* (*Steady Exhalation and Annihilation*), seems to apply. That is, every entity that exists - i.e., everything that contains emp - always and constantly releases "*elementary granules*" (miniphotons, particles, etc.) of opposite parity.

The miniphotons are finally captured by a "*receiver*" and therefore change their dynamics. Since every monad statistically receives miniphotons and others are emitted, the general balance is satisfied.

Logic justifies the exact dependence as $1/r^2$ of both the law of gravitation and the Coulomb law.

Several unexpected implications concern astrophysical and cosmological topics, although an entire critical rethink should only be done when the correct formula for the Doppler effect is available.

We found a "simple" synthetic framework based on "*double helices*", "*single helices*", "*boosters*", dynamics, and antidynamics.

As mentioned above, every monad, which can only be observed from the outside, sometimes reveals - perhaps - characteristics that are reminiscent of the "golden ratio" or fractality. This characteristic should be considered as a premonition of the regularity of the processes which are operative inside the monad - and which we cannot observe. Indeed, if a "Golden Rule" is observed, it is very likely that the unknown internal laws governing the monad must respect the criterion of "additivity" (or self-similarity) and a "variation principle".

Lastly, a fundamental aspect for our understanding of natural phenomena is the nature of the "thermodynamic" approach, which is a strictly necessary requirement.

However, because due to the size of humans, we are unable to directly observe the interactions of individual particles, we must focus on integrated effects over large ensembles of events. But this must be done using correct statistical methods. Finally, natural reality – when interpolated with continuous functions – is intrinsically related to complex conjugate functions, while – due to the empirical constraint – humans can only detect the amplitude, not the phase, of these complex functions.

In any case, all phenomena considered are characterized by double or single, "*complete*" or "*lame*" spiral structures, or by intrinsic characteristics of chirality. In the remaining chapters of our journey, we will see some examples - on very different scales - nevertheless always characterized by specific chirality. Even chemistry or life forms have chirality characteristics, and also phenomena on much larger scales, up to super-clusters of galaxies.

In Chapter 28, we will see how this all leads to a very reasonable hypothesis regarding the endemic origin of life in the Universe. That is, the discussion of the epistemological foundations of physics has led us to surprising conclusions, as they are "beautiful" and "simple", absolutely unexpected. We were surprised by such "beauty" and "simplicity".

The reader, who had the patience to travel this journey, should now consider where we started from, from Rousseau's "*bon sauvage*". If we had then mentioned the conclusions we reached, we would have been completely incredulous, almost out of mind. Instead, everything was derived from "beautiful" arguments, the "simplest" possible arguments. But the surprises are not over ...

Chapter 26

Chirality

"... the scope of [several] *scientists is far away from fundamental problems; their target is to present observational data according to the existing models. They have no time to think, because huge volumes of space data demand a lot of time to process them; ... they have the other target - fast publication of simple papers based on measurements; the success of a project is evaluated by the number of publications, and these rules are very rigid."*

Tamara Kuznetsova (private communication, 2002)

....

Having exhausted the discussion of the basic axiomatic approach, the remaining chapters are dedicated to brief reminders dealing with numerous evidences - which concrete observations confirm, according to what we have considered along the way. In this chapter, we will discuss filament structures (i.e., spirals, or chirality).

Terrestrial atmospheric phenomena

The primary reason for the filament structures lies in Fig. 14.3 here reproposed as Fig. 26.1, of which we also recall the mechanism that explains it.

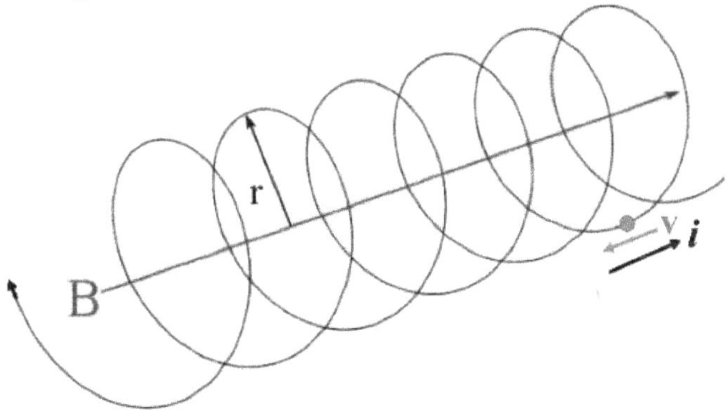

Fig. 26.1. Motion of an electron in an ambient magnetic field.

Every electron - which flows in a region of space where a magnetic field $\boldsymbol{B}$ is present, which in turn is generated by some particle or antiparticle - orients its *element* κ_H perpendicularly to $\boldsymbol{B}$ while its motion along $\boldsymbol{B}$ remains unchanged. The result is a spiral orbit whose radius r depends on the component of the velocity of the electron perpendicular to ***B*** and on the intensity of ***B***.

We therefore only have to explain how ***B*** is generated in space, and the chirality of the phenomenon is thus fully justified. It should be stressed that when it comes to field ***B*** we are referring to the "statistical" treatment (see Chapter 21) which is implicit in Maxwell's equations.

We must now recall a fundamental mechanism, a sort of universal theorem - almost like the Pythagorean theorem or similar - which is *rigorous* and valid throughout the Universe. It provides the explanation for numerous phenomena, many of which are still considered mysterious and unexplained. For example, we start from the smallest phenomena - such as the formation of fog droplets, to phenomena of increasingly larger scale, such as the structure of an electric discharge, or of a cloud, or of stellar and galactic phenomena, up to galactic superclusters. We are referring to what we will briefly call "*Cowling dynamo*" - which derives from the *rigorous* proof of Cowling's generalized theorem that can be stated as follows.

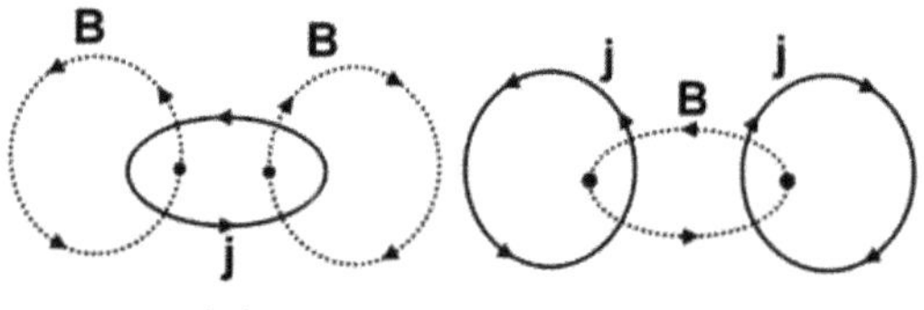

Fig. 26.2. Idealized diagram of every general dynamo consisting of charged matter with some internal dynamics. Only two setups are possible. Case *(a)* has a simple poloidal ***B***, and toroidal electric currents ***j*** or field ***E***. Case *(b)* has a simple toroidal***B*** and poloidal ***j*** or ***E***. The generalized Cowling theorem states that case *(a)* is unstable, while case *(b)* is stable. It is shown that, in both cases, the system reaches the maximum possible efficiency of the dynamo, even if the dynamo is stable only in case *(b)*. But in the case of ideal cylindrical symmetry the energy for case *(b)* is zero. For instance, no cylindrical symmetry occurs when the axis of symmetry in figure *(b)* does not coincide with the system's rotation axes.

Consider electric charges in motion – typically, e.g., an ionized fluid (gas or liquid) moving along closed paths, such as a convective cell. The motion of the charged particles generates a magnetic field ***B***, and an electric field ***E***. The fields ***B*** and ***E*** are linked with each other as required by Maxwell's equations, and the mode of concatenation – or the topology – can only be of two types, shown schematically in Fig. 26.2. Both configurations are equilibrium states (consistent with the approach of traditional physics).

However, the case of Fig. 26.2a is shown to be unstable. In other words, a minimal perturbation is enough for the system configuration to promptly break down and evolve towards a different state (it's a bit like a ball precariously balanced on the top of a hill).

On the contrary, the case of Fig. 26.2b is stable – i.e., the system returns to its state when it is stressed by any perturbation (this is the case of a ball at the bottom of a valley). Therefore, in nature, we will always observe the configuration of Fig. 26.2b.

There is, however, a particular case to consider which - in practice - never finds a counterpart in nature. This is the case of a natural system with ideal cylindrical symmetry. Let's clarify the concept better and consider, for example, one star. It is a huge fireball that rotates around an axis of rotation. Let us consider a section of the star with two half-planes intersecting it along its rotation axis. If the intersections with any two half-planes are identical, the star is said to have ideal cylindrical symmetry. That is, the structure of the star can be obtained by rotating the intersection with a single half-plane by a complete turn of 360° - as it happens, precisely in the generation of a cylinder in elementary geometry. For a system with ideal cylindrical symmetry, the configuration in Fig. 26.2b is the physically possible case. However, it is shown that its energy is zero, i.e., the system can generate no magnetic field $\boldsymbol{B}$ and no electric field $\boldsymbol{E}$. In reality, the case of ideal cylindrical symmetry never exists in nature.

In conclusion, the system – defined in the most general possible way – *always* constitutes a dynamo that generates a magnetic field $\boldsymbol{B}$, and an electric field $\boldsymbol{E}$, which are always linked as in Fig. 26.2b. This dynamo is called the "*Cowling dynamo*".

We do not report the long demonstration (pedantic and scholastic, even if not particularly difficult). However, we think it is worthwhile to give some historical clarification.

In 1892, the American solar astronomer George Ellery Hale (1868-1938) designed a spectrograph that provided images of large areas of the Sun using only a single spectral line. In 1919-1920, the Irish physicist and mathematician Sir Joseph Larmor FRS (1857-1942) proposed the existence of a dynamo inside the Sun and inside every star. Today we know that the suggestion was correct, given the enormous endogenous energy generated inside every star by thermonuclear fusion reactions.

In 1933-1934 the English astronomer Thomas George Cowling (1906-1990) published his famous theorem which asserted that any dynamo, generated by a system with *perfect cylindrical symmetry*, can generate no magnetic field $\boldsymbol{B}$. The proof was based on reasonable approximations. A heated debate followed. New proofs followed – always approximate – but the result was always the same - and we point out how this result is perfectly consistent with the aforementioned more general result.

Since then, Cowling's theorem became a nightmare for solar and stellar astrophysicists. They just didn't know how to get out of this deadlock. They implemented stellar dynamo models knowing *a priori* that they could not work. The hypothesis of *perfect cylindrical symmetry* was devastating and they did not realize it.

Meanwhile, in 1955 the two American astronomers Harold Delos Babcock (1882-1968) and his son Horace Welcome Babcock (1912-2003) managed to do what Hale had not managed, and drew detailed maps of the $\boldsymbol{B}$ field of the visible surface of the Sun.

In 2002 - almost without realizing it - Giovanni P. Gregori proved, rigorously, without any approximations whatsoever, the *generalized Cowling theorem* mentioned above. The nightmare was over - but much more: the rigorous theorem holds on every spatial scale from a fog droplet up to galactic superclusters. This is what we try to document very briefly here.

We begin by considering the mysterious process of *atmospheric moisture that agglomerates to form a droplet.* It is well known that it is indispensable to have a condensation nucleus. This can be a tiny speck of dust, or even a simple ion, or even one of the particles that always rain down from space above the atmosphere and are known as cosmic rays. However, it could also be a simple ion that is formed by rubbing when the air - moved by wind - slides on the terrestrial surface (of lands or oceans etc.).

All sources of condensation nuclei are effective. A water molecule (H_2O) is an electric dipole, i.e., it is similar to a bar with one end charged with a positive electric charge and the other end negatively charged. Therefore, around a condensation nucleus, a first layer of H_2O molecules will be formed, and the new agglomerate will have the same total electric charge as the initial concentration nucleus (Fig. 26.3). A new layer of molecules of H_2O etc. will thus be formed. The process is well known, but it is much, much too slow to explain why a real droplet can form.

This is where the role of Cowling micro-dynamos comes into play crucially. Tiny temperature differences in the air generate convection micro-cells that generate small $\boldsymbol{E}$ -fields. Agglomerates of H_2O that have an electrical charge move due to the $\boldsymbol{E}$ field, while H_2O molecules that are electrically neutral are not affected by the field $\boldsymbol{E}$. That is, every agglomerate moves in the environment saturated with H_2O molecules and collects them just as a broom collects rubbish. Thus, an actual droplet is formed, which can remain suspended in the air or fall depending on its size, etc. This is the only explanation known today that can explain why atmospheric precipitations occur.

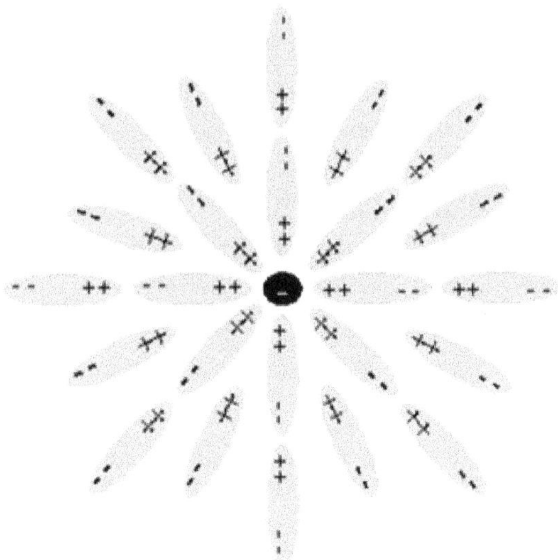

Fig. 26.3. Qualitative, out-of-scale sketch explaining the agglomeration of water molecules as electric dipoles around a condensation nucleus with a negative electrical charge. A symmetric case occurs for a condensation nucleus with a positive electric charge. See text.

Let us now consider the case of convective cells measuring a few tens of centimeters, such as are sometimes formed, e.g., inside domestic chimneys or fireplaces, or inside the cockpits of aircrafts. A phenomenon is born - which is rare but not too rare - and a luminous globe is formed, known by various names. We will adopt the more common name, "*ball lightning*". These phenomena last a few dozen seconds and then disappear with a slight noise. They can pass through window glass, or even through the cockpits of airplanes, generally without leaving any consequences other than an easily imaginable scare for the unaware passengers. The phenomenon has also been reproduced in the laboratory and is generally denoted as *BL* (ball lightning).

The Cowling dynamo generates a toroidal field (i.e., donut-shaped, as in Fig. 26.2b) which confines the hot ionized air inside it, which reaches a few thousand degrees centigrade, therefore making the air luminous. But the *BL* surface is surprisingly virtually cold. The phenomenon lasts as long as the internal thermal energy of the *BL* is sufficient to maintain the convective motion, which is the basis of the Cowling dynamo.

Let us now consider the physics – currently believed to be little known – of what happens in a normal electric spark, or in an ordinary lightning bolt. We wonder why the electric discharge occurs according to a linear, zigzag structure, instead of not involving an entire large section of space. The phenomenon can be explained very easily by considering that the speed of propagation of the electric discharge - measured and much lower than the speed of light - agrees well with a domino effect process, whereby a *BL* in the act of extinguishing ignites an adjacent *BL*, etc.

When we consider larger scales, convective motions in clouds have the size of a cloud. Thus, large Cowling dynamos generate ***E*** fields which cause electrical discharges, mainly cloud to cloud, much more frequently than cloud to ground. These ***E*** fields also determine the positive electrical charge of the upper atmosphere, in addition to the *TLE* phenomena (transient light events), which are a sort of lightning that occur above the clouds up to the upper layers of the atmosphere (ionosphere). These are all well-known phenomena - even if almost all of them are either not explained, or are only partially explained. It is not possible to go into any further detail here.

We will instead remember the mysterious *TGF*s (terrestrial gamma ray flashes), discovered by astrophysicists while looking for gamma ray sources coming from space - and instead found, with much surprise, phenomena coming from the upper part of the clouds. It should be noted that this phenomenon is not purely academic, given that it is considered very probable that many air crashes - apparently unexplained - have been caused by the aircraft having hit a *TGF* which instantly knocks out all the electronic circuits for the control of the aircraft and communications, including black box circuits, etc. The topic is not only of academic interest, but of tragic concern.

Astronomy phenomena

In our quick exemplary overview, we now consider the evaporation which - due to the enormous endogenous temperature - occurs permanently from a star or from the Sun. This is the case of the solar wind, described as "expansion of the solar corona". Strangely, according to very careful studies, it has been assessed that the solar wind is inexplicably self-confined, i.e., contained in thread-like structures, much more than one should expect. The explanation lies in the small thermal non-uniformities intrinsic to the solar wind. These generate a myriad of small Cowling dynamos that produce solenoidal ***B*** fields that determine a self-collimation process.

But the phenomenon of filamentary structures is regularly observed in many images of the cosmos (e.g., see the splendid Fig. 26.4).

When a star is forming - or even when the ***B*** field of a galaxy is considered - the reference structure is summarized by the sketch in Fig. 26.5. Note a toroidal (donut-shaped) structure that appears to be due to the ***B*** field, while the ***E*** field should be represented by the two violent jets, which in fact are a frequently observed phenomenon. This structure is called the *AGN* (active galactic nucleus).

We show some real images from a few years ago provided by *NASA* and other space agencies (*ESO* etc.), while *NASA*'s *James Webb Space Telescope* is now providing frequent images of breathtaking beauty. Fig. 26.6 is perhaps a classic example by now. Fig. 26.7 is a mosaic of galaxy images released by *NASA* in 2013, obtained with the *Hubble Space Telescope*.

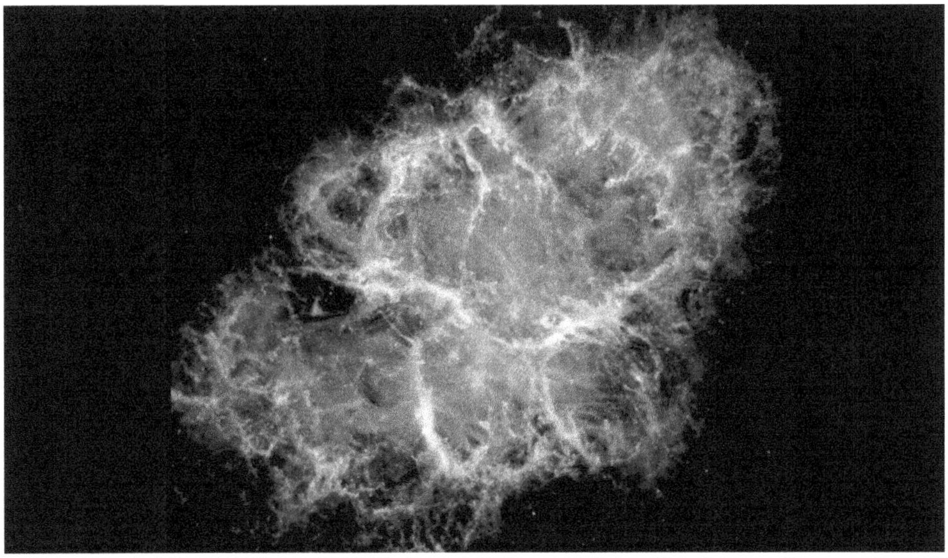

Fig. 26.4. *NASA*'s *James Webb Space Telescope* has sliced open the structure of the Crab Nebula, helping astronomers as they continue to evaluate leading theories about the origins of the supernova remnant, or an exploded star. (Image credit *NASA, ESA, CSA, STScI, Tea Temim - Princeton University*).

Fig. 26.5. Sketch showing a model of an *AGN* that recalls the general structure and logic of the interpretation of various astrophysical observations, explained as follows. A central supermassive black hole is shown by a small black circle. The colored bands are believed to indicate regions responsible for the emission of "narrow" optical lines. The sketch is taken from the poster of the meeting "*Gamma-ray bursts in the afterglow era*", 3rd workshop, Rome, CNR Headquarters, 17-20 September 2002. The Proceedings were published by the *Astronomical Society of the Pacific* (*Conference Series*, book 312) and a similar black and white sketch is shown on the cover.

Figure 26.6. The Crab Nebula seen in X-rays is a real observed case of a galactic ***B*** field resembling an *AGN*. The Crab Nebula in the visible and UV cannot provide evidence for the toroidal model, Image *NASA/CXC/Smithsonian Astrophysical Obs.*).

Finally, let us briefly remember the well-known images of spiral-shaped galaxies. The explanation usually given is based on gravitational attraction. However, it seems strange that all galaxies show spiral branches roughly lying within one plane.

Instead, this phenomenon is well explained by considering that every central nucleus of a galaxy clearly shows a toroidal ***B*** field. The κ_H *elements* of what implodes on the central nucleus tend to orient themselves in the plane of the "donut" of the toroidal ***B*** field. So every evolving galaxy will tend to align approximately with the mean plane of the ***B*** field of the central galactic nucleus – which in turn is determined by a huge Cowling dynamo.

Finally, let us recall - just to mention - two phenomena that have been clearly observed.

The stars within the galaxies appear mysteriously located along elongated structures, like imaginary filaments. The same curious filament structure is also observed with reference to the galaxies that are part of a supercluster of galaxies.

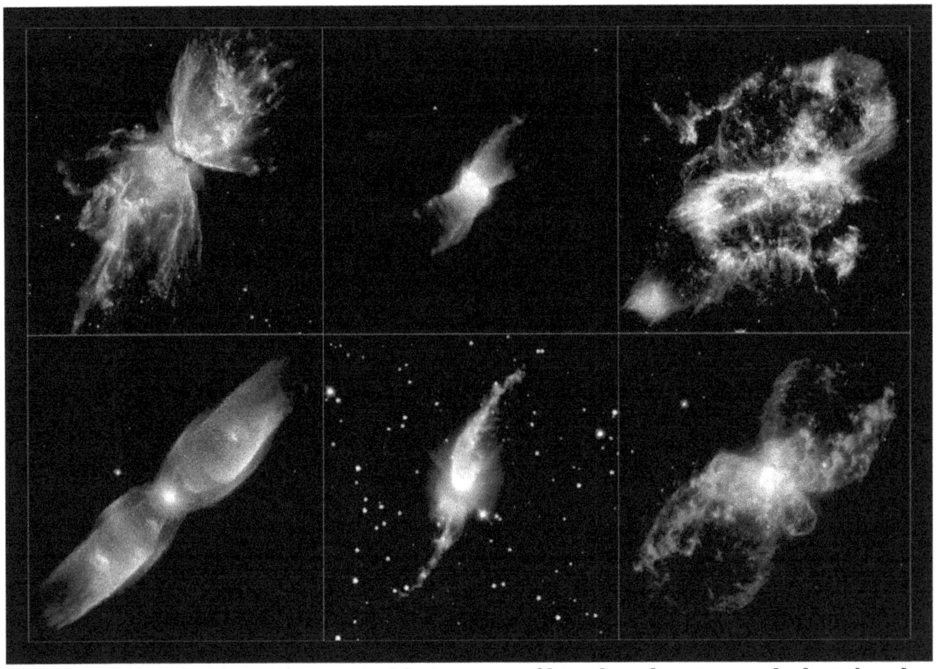

Fig. 26.7. Selection of remarkable images of bipolar planetary nebulæ taken by *Hubble*. The galaxies are (left to right, and top to bottom). Line 1: NGC 6302, NGC 6881, NGC 5189. Line 2: M2-9, Hen 3-1475, Hubble 5. Image published 4th September 2013. Credits: *ESA/Hubble* and *NASA*; Bruce Balick (*University of Washington*), Vincent Icke (*Leiden University*, the Netherlands).

It seems that no explanation is given for these seemingly very curious phenomena. Indeed, this is the consequence of enormous processes due to galactic-scale Cowling dynamos, and galaxy superclusters, respectively. The filament structures would therefore be due to a phenomenon similar to that which determines the self-collimation of the solar wind or stellar wind (shown in Fig. 26.4).

Molecular phenomena

Molecular bonds cannot be treated with Maxwell's equations, and thus with Cowling's dynamo.

It is necessary to refer to the interactions between *elements* κ_E e κ_H of the volleys of electrons associated with different atoms. The result is the wide variety and diversification of chemical bonds, both inorganic and organic chemistry. The resulting intricate structures are well known, all characterized by spiral shapes. The best-known case is the spiral structure of *DNA*.

It would take too long here to try to briefly summarize a phenomenology that is very varied and important for the formation and development of life. It is enough for us to have recalled here this fundamental aspect of natural reality.

Chapter 27

Antifragility and Autopoiesis

"The secret to happiness is freedom …
And the secret to freedom is courage."
[Thucydides (c. 460 - c. 400 BC)]

………………………………………

The concept of antifragility is drawn from statistical studies related to financial applications and was introduced by Nassim Nicholas Taleb (1960-), a Lebanese-American mathematical statistician and essayist. His studies concern chance, probability, and uncertainty.

Taleb is the author of "*Incerto*", a five-volume philosophical essay on uncertainty published between 2001 and 2018 (including "*Antifragile*", which is the 4th book). He has been a professor at several universities, having served as *Distinguished Professor of Risk Engineering* at the *New York University, Tandon School of Engineering* since September 2008. Since September 2014, he has been co-editor-in-chief of the academic journal *Risk and Decision Analysis*. Among his many accolades, *The Sunday Times* wrote that his 2007 book "*The Black Swan*" is one of the *12* most influential books since World War II.

The concept of antifragility is based on examples from science and mathematics and states that some systems are strengthened by encounters with disorder. The author wants to connect disorder to nonlinear responses and optionality. Similar to every highly innovative concept, his insightful analysis has sometimes been debated and controversial. In any case, the concept of antifragility focuses on a relevant concrete aspect of reality that has not been previously observed and - in fact based on real observations - fits very well with the discussion in our journey. The relevance of this concept to our general axiomatic aspect was highlighted by Bruno G. Gregori, after reading "*The Black Swan*" and "*Antifragile*".

The concept of antifragility is linked to the ability of nature to replicate its structures. The definition of antifragility is perhaps best expressed by referring to mythology, as Taleb suggests.

"Fragility" is symbolized by the myth of the *sword of Damocles*. On the contrary, the ideal ability to restore the previous state is symbolized by the myth of the *Phoenix*. In physics, this is equivalent to perfect "elasticity" and "reversibility". But the ideal "elasticity" is a simplifying (non-physical) abstraction by human mind, i.e. an abstraction that, however, implicitly rejects the possibility of explaining the irreversibility and the one-way direction of the arrow of time (see Chapter 9 and *passim* elsewhere).

"Antifragility" is, rather, expressed by the myth of the *Hydra*. The *Lernæan Hydra*, or simply *Hydra*, was a serpent-like aquatic monster from ancient classical mythology. In the canonical myth, the monster was killed by Heracles as the second of his Twelve Labors. The *Hydra* had many heads. According to later versions, the *Hydra* had a regenerative property, so that for every head cut off, the *Hydra* grew back two. Heracles managed to cut off all the heads and burn their necks using a sword and fire.

In fact, similarly to antifragility, "statistics" must take into account complexity or chaos (see Chapter 18), properties that are both the opposite of every linear approach, i.e., the opposite of any traditional scheme based on the search for chains of cause and effect. The cause-effect scheme is more suited to the requirements of "simplicity" required by the limitations of the human mind and by the Galilean formulation. That is, complexity, or chaos, and antifragility are essential and fundamental aspects of our cognitive process, and these aspects are of concern for all the "monads" that we can observe from the outside (see Chapter 6), including organic molecules or living beings - even if these aspects are formally excessively complicated for conventional mathematical treatment.

The difference between complexity and antifragility must be emphasized. The fundamental laws of "statistics" are different in different domains (see Chapter 18) and are generally unknown. Antifragility is similar and more general than complexity, in that it can be applied to the world of living matter. In fact, the same logic of antifragility leads to a reasonable hypothesis about the origin of life, evolution, the ability of living beings to restore damaged parts, and finally also to the survival strategy of generating new individuals to avoid the inconvenience of aging, etc. Ultimately, this is the key to survival in terms of regeneration of "living" beings, since this is the only possible way to circumvent the inconvenience of the unidirectional direction of the arrow of time or the constant increase of entropy (see Chapter 9).

Antifragility is a system's ability to regenerate a damaged component. For example, a wound, a broken bone, etc., heals thanks to the natural regeneration ability of the animal or human body. The most obvious example is vegetation. The grass regrows after every cut, or the pruning of a tree promotes the development and productivity of the remaining branches. Or even, e.g., in the case of an earthworm, if some segments of the earthworm's tail are torn off, in some species the segments simply regrow, even if earthworms never regenerate a greater number of segments than they lose. Apparently, every segment produces a tiny electrical charge, and the earthworm regenerates the missing segments until it reaches its maximum expected electrical charge again. That is, notice how electromagnetic processes always appear to play a crucial role at the molecular level.

Another example is the *Linckia* starfish, which is able to regenerate entire new individuals starting from an arm, plus a piece of the central body that has detached from the rest of the organism. Thus, a single dismembered sea star can potentially produce up to five new individuals via asexual reproduction. A recent study by *Stanford University* used genetic and molecular tools and solved the mystery. In fact, the "head" is distributed in multiple regions, including the center and every limb. This finding, therefore, raises serious questions about traditional concepts and suggests a complex evolutionary history. It is important to study different life forms to gain information about evolutionary biology.

But "antifragility" also applies to other life forms that are always very elementary. For instance, consider the case of the so-called brittle stars, or serpent stars or *ophiuroids* - from the Latin *ophiurus* "brittle star"; from ancient Greek ὄφις (óphis) "snake" and οὐρά (ourá) "tail", referring to the serpentine arms of the fragile star. Brittle stars are echinoderms of the class *Ophiuroidea*, closely related to starfish. They crawl along the seabed using their flexible arms for locomotion. They typically have five long, thin, whip-like arms, which in the largest specimens can reach up to 60 *cm* in length. The *Ophiuroidea* include the *Ophiurida* (brittle stars) and the *Euryalida* (basket stars). Over *2,000* species of brittle stars are known today, and more than *1,200* of these are found in waters deeper than 200 *m*.

Ophiuroids have no brain, but a recent study has shown that they can learn through experience. They use their nerve cords to learn by association – a concept proven in classical conditioning.

Another example is the *Cladonema* jellyfish, which can regrow its tentacles in just a few days, highlighting the role of unique proliferative, stem-like cells in this rapid regenerative process.

But the same concept also applies to recent regenerative studies for damage sustained by the human nervous system. For example, we will remember only one recent study (2023) - among many - by the *University of California* (*Los Angeles, Health Sciences*) and by the *École polytechnique fédérale* of Lausanne. They have developed a gene therapy - which has been demonstrated in mice - to stimulate the regrowth of nerves through spinal cord injury and guide the nerves to reconnect. When the spinal cords of mice and humans are partially damaged, initial paralysis is followed by extensive and spontaneous recovery of motor function. However, following a complete spinal cord injury, this natural spinal cord repair does not occur, and there is no recovery. Significant recovery after severe injury requires strategies that promote nerve fiber regeneration – but the conditions that are necessary for these strategies to successfully restore motor function have remained elusive. Scientists performed extensive analyzes to identify which type of neuron is involved in natural repair. They thus proposed a design for a multifaceted gene therapy. They activated growth programs in the identified neurons in the mice to regenerate their nerve fibers. They say they took inspiration from nature when they designed a therapeutic strategy that replicates spinal cord repair mechanisms that occur spontaneously after partial injuries. Let's not go into these and other details.

In summary, antifragility is a peculiar characteristic of the "statistics" that governs the world of living matter. That is, in principle, "antifragility" can be different from the usual way of thinking about complexity and chaos. The concept is analogous to - although more general than - complexity, according to some criterion that at the moment is unknown to us. Therefore, a concept of "generalized" chaos can be introduced, since this denotes a form of "chaos" that is based on some new laws of "statistics" that are not yet known to us.

According to the logic of our logical journey, it is probable that antifragility can be explained by means of the different interaction between κ_E and κ_H *elements* of the volleys of electrons associated with different atoms. That is, the filamentary structures are generated through the interaction of κ_E and κ_H *elements* and the chiral models are generated through the interaction between κ_H *elements* (see Fig. 26.1). The innovative content of this "insight" can be appreciated, e.g., by considering some very recent studies. We emphasize, however, that these investigations are based on linear phenomena, i.e., they presuppose a cause-and-effect chain.

In contrast, antifragile systems take advantage of chaos to enhance their life performance. This is a constant process of ongoing innovation. This explains the evolution and ability of living forms to adapt to a changing environment. In other words, every cause-and-effect chain is related to linear processes and fragility. Chaos, evolution, and complexity are, in a certain sense, better correlated with antifragility.

But Nature goes further still. Everything evolves over time, everything ages, entropy increases inexorably (see Chapter 8) - and this is inherent in the very perception of irreversibility and therefore in the perception of time, of aging, and of the implicit "race" towards death. The perception of irreversibility is common to all living forms, of all levels, precisely because of a problem of survival - even if, perhaps, only man retains memory of it. Nature then came up with an antidote for survival. This is autopoiesis, i.e., the ability of natural systems to generate new “young” individuals who, upon the death of “aged” individuals, take their place. Without this basic property, any existing system would be doomed to extinction, with no possibility of exception.

Our considerations, therefore, led us to consider the role of the formation of life in the Universe, and of its evolution and adaptation to environmental vicissitudes. This is discussed in the next chapter.

Chapter 28

Life and Antilife in the Universe

"Humanity - Humanity is a great machine, perfect in its imperfection. The fuel that drives it is intellect, *while the button to switch it on is* faith. *Humanity is a wide galaxy, whose center is made up of* intelligence, *while the stars represent* hope. *Humanity is a long narrow road, made of obstacles and satisfaction, of ups and downs. We know not where it leads to, but we imagine, dream, and so the destination is not the end of the journey, but it's the beginning. Humanity is all these descriptions, one only cannot be chosen, but they can be mixed together.*

Humanity is a great machine, the center of its own galaxy, which follows a road towards its uncertain future. Humanity, one always wonders what it is, who we are? What we want to obtain? Why? How? When?... Man has always wanted to explore outside himself, forgetting to explore inside himself; sooner or later, I'm absolutely sure, he will find the maturity or simply the courage to do so."

English translation kindly made by Susan Brodar,
who is warmly acknowledged.
Francesco Spadoni (2002-2024)
[composed on 15th April 2016, private communication, 2016]

....

In our logical journey, we began with truly minimal hypotheses, aimed only at taking into account the essential information provided by experiments, trying to interpret them with the minimum possible number of arbitrary assumptions, with the sole concern of avoiding the paradoxes that have afflicted theoretical physics.

Step by step, we have thus arrived at a set of "discoveries" that we could never have imagined *a priori*, and that have often truly surprised us. But the amazement is not over yet, given that we can give a reasonable explanation for the appearance of life not only on Earth, but in the entire Universe.

But let's go in order. First, we formulate the final thesis, then we discuss the evidence that concrete observations on Earth can provide to justify the entire explanation that we give of this disquieting and fascinating topic.

The thesis in our journey asserts that life is an endemic phenomenon in all the matter of galaxies – just as antilife is an endemic phenomenon in the antimatter of antigalaxies. Life or antilife can reach different degrees of evolution or diversification, depending on environmental conditions, which can be more or less favorable to the development of certain forms instead of others. Abrupt environmental changes can cause some species to become extinct, while others develop and take over.

As for the evidence we can find on Earth, we will see how life is continuously self-regenerating on the deep ocean floor, fueled by the only source of energy available there, i.e., the endogenous heat exhaled from seafloor.

We will therefore first discuss what we know about the endogenous heat exhaling from the ocean and sea floor. Then, we will consider what we know about life in the deep ocean. Finally, we will discuss what similar phenomena can be expected to happen on other celestial bodies for which we have some information given by interplanetary probes.

The endogenous heat

A very widespread belief - yet unmotivated and unproven by anything - conceives the Earth as an initially very hot ball that cools over the course of several billion years. The age of the Earth - and of the entire Solar System - is estimated to be around $4.5 - 4.6$ *Ga* (*Giga-years*, or billions of years). The famous French naturalist Georges-Louis Leclerc, Comte de Buffon (1707-1788) - known as Buffon - carried out experiments on the cooling of cannonballs, and - although it seems that no one remembers Buffon anymore - it seems that the establishment agrees with this hypothesis as unanimous, even if no experimental evidence can confirm it.

The reader should therefore forget this "legend" - and consider instead that the Earth is not a rigid body like a rock, but has various components more or less strongly linked together. Also consider that the gravitational action is very intense and is exerted on the body of our planet Earth both by the Sun and the Moon. What is of interest here, however, is not so much the total intensity of the gravitational force, but rather the difference between the action exerted in different points of the Earth. That is, what matters is the difference between the gravitational action exerted on the side of the Earth facing the Sun (or the Moon) and the action on the opposite side of the Earth. In this way, the gravitational action determines a relative sliding of different components of the Earth's body. In mathematical language, it is said that what matters is the "gradient" of the gravitational force. In this case, the influence of the Moon - due to the gravitational gradient - is much more important than the influence of the Sun, and the most obvious result are the ocean tides. However, the tides of the solid Earth are also very important, and one of the most evident effects is the slow and steady lengthening of the duration of the day (*l.o.d.*, length of the day).

But we will not discuss these fascinating and non-essential, aspects for our journey. Rather, we will consider the fact that the different components of the Earth – mobile relative to each other – are ionized, i.e., electrically charged. Therefore, owing to the drag and mutual movement due to the gravitational gradient, the entire Earth system operates like a true dynamo. We use the acronym *TD* – dynamo, for "tide-driven". After a very long study of the literature, perhaps lasting a few decades, in 2002, Giovanni P. Gregori published the energy balance of the terrestrial phenomena associated with it. The outcome is shown in Table 28.1 and Fig. 28.1. Note the absolutely surprising data for the intensity of power (energy produced per unit of time), because it can explain *all* phenomena connected to the Earth's endogenous energy. It is also very surprising that the largest percentage value is associated with the flow of geothermal heat, which far exceeds the power absorbed by volcanic and seismic phenomena.

Table 28.1. Endogenous energy of the Earth during the last million years ($\times 10^{13}$ W)

Type of power released	estimated value	rounded value	percent
global heat flow measured at Earth's surface	4.42 ± 0.10	~ 4.42	~ 79%
tidal friction (total)	$0.4 - 0.7$	~ 0.55	~ 10%
magmatic heat flow	$0.20 - 0.39$	~ 0.3	~ 5%
metamorphic heat for transformation of sedimentary and volcanic rocks	of the same order of magnitude	~ 0.3	~ 5%
gravity sliding	2.5×10^{-2}	~ 0.025	~ 0.5%
seismic	$0.3 - 3 \times 10^{-3}$		~ 0.5%
potential energy loss due to soil erosion	3×10^{-4}		~ 0.0%
Total		~ 5.56	100%

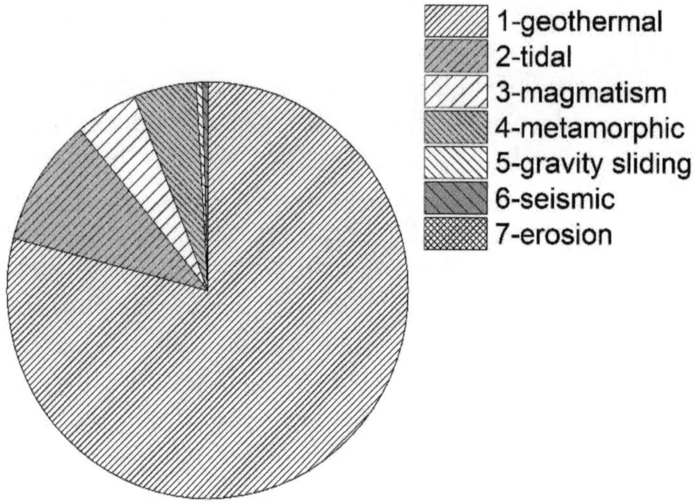

Fig. 28.1. Relative role, during the last million years, of different types of endogenous energy manifested in Earth's phenomena (see Table 28.1).

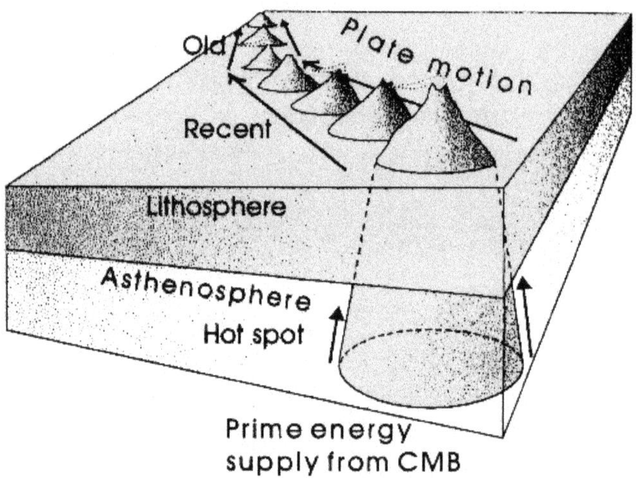

Fig. 28.2. The chain formed by the Hawai'i Islands and the Emperor Seamounts is composed of *108* elements (islands or seamounts) of increasing age in sequence from Kīlauea towards the Kamchatka Peninsula. Their classical interpretation (according to John Tuzo Wilson, 1908-1993, Canadian geologist and geophysicist) is due to the drift of the lithosphere, which slides analogously to a conveyor belt that collects the magma erupted from a fixed hotspot. By measuring the volume of the elements of the chain and their age, it is thus possible to monitor both the temporal variation of the magma eruption and the speed of the drift of the lithosphere. Note the discrete sequence of individual elements - instead of a continuous belt of erupted magma - due to the role of every volcanic edifice acting as a safety valve, like in a pressure cooker. The data interpolation is based on the measured age and on the position of the elements of the chain.

This same mechanism is the explanation for the generation of the magnetic field - both of the Earth and of every other planetary object smaller than a star.

For the sake of brevity, we cannot insist here on the multiple implications of this very surprising result. Let us rather recall another no less surprising piece of evidence, which concerns what can be called the Earth's electrocardiogram.

Let us remember that - as has been well known for several decades - the lithosphere of the Pacific Ocean flows, like a treadmill, over the hotspot of Hawai'i, that in this way has generated a sequence of *108* islands and seamounts (Fig.28.2).

Let's remember the definition of lithosphere, which is the upper shallow layer of the Earth with a thickness between a few tens and a few hundreds of kilometers, while the Earth's crust is a much shallower layer than the lithosphere. The lithosphere is characterized by a base that rests on a layer with very low friction, almost as if it were lubricated, allowing easy sliding. This sliding is shown by continental drift.

Combining the position and volume of everyone of the *108* islands and seamounts, together with the date (when available) of the oldest local lavas, the result of the interpolation is shown in Fig. 28.3, which shows a trend reminiscent of an electrocardiogram. One beat occurs every $\sim 27.4 \pm 0.05\ Ma$ (*millions of years*). For what follows, it must be emphasized that it is impossible to reduce the uncertainty to less than $0.05\ Ma$, considering the type of available observational information. Every beat lasts a few million years, and corresponds to the formation, somewhere on the terrestrial globe, of one or more large igneous provinces (*LIP*s). Humans are experiencing a heartbeat for the first time in their existence. The *LIP* corresponding to the current heartbeat is likely Iceland, which did not exist $\sim 2\ Ma$ ago

The two plots in Fig. 28.3 show, respectively, from top to bottom, the volume of magma erupted per unit time, and the speed of the lithosphere on the treadmill with respect to the endogenous *hotspot*. The endogenous *hotspot* releases heat producing lava which, over the course of approximately $70\ Ma$, generated the *108* islands and seamounts. Note that the speed estimate is likely to be more reliable, as it is less affected by errors in the calculation of the volume of every volcanic edifice.

It should also be stressed that, according to investigations not reported here in detail, it appears that the friction at the base of the lithosphere is greatly reduced during a heartbeat, given that the base of the lithosphere is very heated. During the remaining long beat-to-beat time intervals of the electrocardiogram, the friction becomes very large (the average is 96.34% viscous coupling, or lubricated state, and 3.66% elastic coupling, i.e. rigid coupling state, with enormous excursions, although the estimated errors can be of same order of magnitude as the effect that needs to be detected). In any case, a substantial reduction in friction is a harbinger of a high sliding drift of the lithosphere, with consequent enhanced geodynamic, i.e., seismic and volcanic, activity and uplift of mountains, etc.

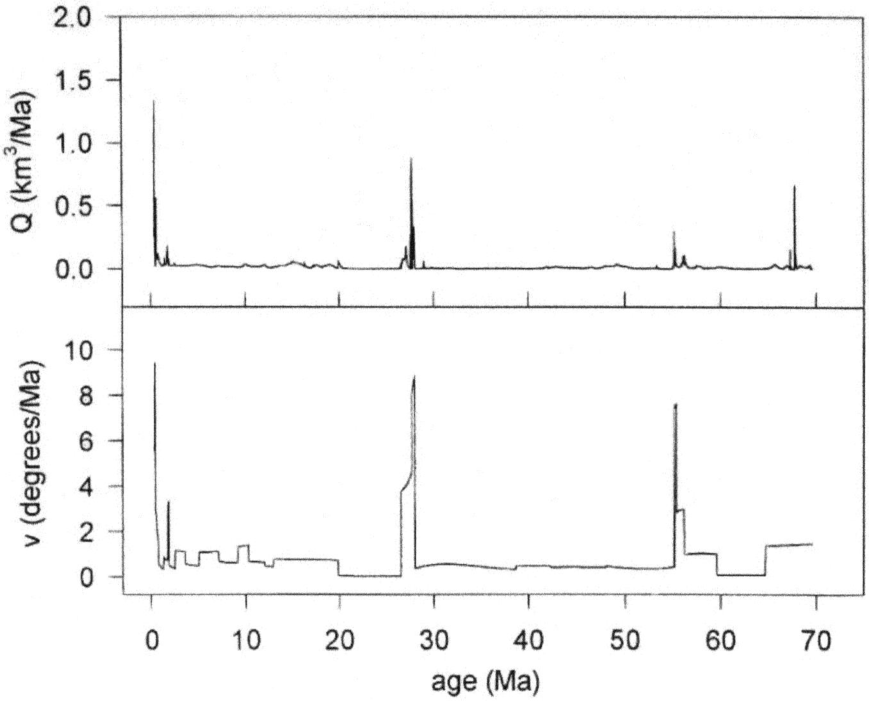

Fig. 28.3. Electrocardiogram of the Earth - or variation per unit time of magma eruption from the Hawai'i hotspot during the last $\sim 70\ Ma$, obtained by formal interpolation over the volume from the Hawai'i Islands and the Emperor Seamounts with reference to their respective ages. The resolution time is $100.000\ anni$. The pattern is reminiscent of an electrocardiogram. A heartbeat occurs every $\sim 27.4 \pm 0{,}05\ Ma$, a trend that appears to have been a steady feature during the last $\sim 250\ Ma$, or perhaps even much longer. The upper plot shows the amount of magmatic eruption in km^3/Ma. The lower plot shows the drifting speed of the lithosphere expressed in degrees per million years (one degree corresponds to approximately $110\ km$ on the Earth's surface). Thus, a great variability is found, showing peaks reaching the enormous speed of $3\ mm/day$. See text.

However, the most dramatic impact concerns climate change. A considerable increase in the input of endogenous energy into the oceans/atmosphere natural system leads to violent climate changes, which in fact we are all witnessing today. Nor it is possible for us to know if we are - now - already at the peak of the beat, or if the peak of change will still have to grow.

On the occasion of every heartbeat, mainly during the most intense heartbeats, dramatic extinctions occurred of species – beginning from life forms living in the oceans, while other species took over. The concern about the current heartbeat is therefore to contain and mitigate its effects, and to prevent the human species from being among the next species to go extinct.

All this analysis was published in 1996 by Giovanni P. Gregori in cooperation with Prof. Dong Wenjie, a very astute scientist - as well as an excellent friend - who was spending a sabbatical in Rome, and is now *Professor* and *Dean* of the *School of Atmospheric Science*, at *Sun Yat-sen University*, Zhuhai, Guangdong, (PRC), as well as Director of the *Future Earth Global Secretariat Hub China* and Secretary General of the *FE Chinese National Committee*.

It would take too long to explain here the reasons which determine this particular electrocardiogram-shaped trend of Fig, 28.3. In any case - in summary - the release of endogenous heat is a leading driver for climate control, and also for the development and/or survival of life forms. Climate cannot be considered as a very slowly evolving phenomenon, such as that of the unproven hypothesis by Buffon's cannonballs.

Life in the deep ocean

An absolute leading figure for the discovery of life in the ocean depths is Prof. Martin Torvald Hovland, a Norwegian scientist of rare acumen and enthusiasm, a true scientist who "reads the book written by Mother Nature". Trained as a meteorologist physicist, in 1976 he was hired by *Statoil* - a new expanding Norwegian oil exploration company. Prof. Hovland had the task of explaining the seabed to underwater engineers through geophysical mapping. In 1985, *Statoil* had a tragic accident while drilling a well off the coast of central Norway. An explosion occurred due to shallow gas that was penetrated by the drill from a floating drilling platform. It was impossible to control the return flow of gas, sand and water. The floating platform had to be abandoned, and one casualty occurred. A task force was then established aimed to find a way to predict where surface gas is located in the subsurface sediments of seafloor.

Given Hovland's interest in "pockmarks" (sort of spots on the ocean floor that resemble the scars left by chickenpox) and surface gas, Hovland was chosen to lead this group of five scientists. Hovland thus had the opportunity to cooperate with universities, and with highly qualified specialists in every discipline of Earth sciences, as well as to take advantage of means of experimental investigation - operated all over the world - means that no research institute could ever have had available.

His exceptional enthusiasm - worthy of a teenager - added to his acute intelligence, culture, and versatility, allowed him to obtain a large quantity of new discoveries, which are not yet fully acknowledged by the official academic environment. As always happens, every innovator who makes substantial as well as incontrovertible discoveries is considered a subverter of the pre-established establishment, and viewed with instinctive distrust. We certainly cannot list here all the huge contributions due to Prof. Hovland.

Among other discoveries, Hovland demonstrated how on the ocean floor, at a depth of a few kilometers - where sunlight cannot reach - life continuously regenerates as a result of the exhalation of methane (CH_4) from the ocean floor, which transports endogenous heat. So it happens that very basic life forms, microorganisms, are the beginning of the food chain. White, eyeless crabs compete for mud more or less rich in these microorganisms. Increasingly advanced forms of life develop. The best-known forms are long cylindrical structures, white with an orange upper segment, which rise from the ocean floor for a few meters, where intense heat flows occur. This phenomenon occurs predominantly on the crest of ocean ridges. Images have been reported for several years, taken by usually unmanned robots.

A very recent discovery[3] reports gigantic worm-like bodies, discovered within cavities in the subsoil of these sites on the crest of the oceanic ridges, which would seem to be analogous to the columnar structures already known and mentioned above, and which rise above the ocean floor.

Corals are another related phenomenon, which develop where sea currents transport water rich in methane (CH_4) and other hot gases that exhale from the ocean floor.

The biological and food chain that is thus generated is increasingly advanced, and progressively occupies shallower layers of the oceans, until it finds a new very important source of energy, namely solar radiation. Life can thus explode in all its wonderful manifestations, to the point of occupying the emerged lands. This discovery was both fascinating and striking.

---

[3] See https://www.livescience.com/animals/ghostly-white-giant-worms-appear-to-be-reproducing-under-the-seafloor-where-tectonic-plates-meet?utm_medium=referral&utm_source=pushly&utm_campaign=Animals.

In this same sense, highly evolved life forms have been surprisingly observed with remote-controlled robots in the canyons near the oceanic ridge of the Chilean coast,[4] in the vicinity of large methane sources.

In this same respect, we recall a result previously acquired by the Russian sedimentological school - as reported by Prof. Aleksander Borisovich Ronov - who years ago had defined the "*principle of life preservation*". On the basis of the volumes observed globally of various types of rocks, it is asserted that gas exhalations from the soil are essential for the evolution and survival of life. That is, with no CO_2 and with no CH_4 exhaling from the soil, life would become extinct due to a lack of carbon source. Hovland showed the phenomenon in detail, almost in real time.

Without going into further details of this fascinating discipline, let's now turn our attention to the life forms in the depths of the largest ocean trenches. This little-known aspect of Nature is crucial for understanding the realistic possibility of life on other planetary bodies (see below).

For the sake of brevity, we leave out everything related to microorganisms, or any very basic life form that is found in these ocean trenches. Let's ask instead what we know about animals of considerable size.

Let us also briefly mention the phenomenon of beachings of giant squid (weighing several quintals), which are occasionally reported from different parts of the globe.

Normally, these beaching all occur within a few months all over the world, and are then not reported for long periods of time. It would seem that - consistently with the explanation given for the great extinctions mentioned above - sudden, occasional, and temporary increases in methane (CH_4) emissions from the ocean floors all over the world make the water polluted. These giants therefore search for purer waters and move to much shallower depths, until they die from embolism. Furthermore, we do not know the depth at which these beings live normally.

Let's remember instead that scientists have dropped free-fall traps with mackerel baits and equipped them with cameras. The traps sink using a steel ballast weight, and it can take around $4\ hours$ for the trap to sink to the bottom of the ocean trench. The traps and cameras then remain at depth for around $12 - 24\ hours$. In this way, they collected specimens of an unusual creature, which they called snail fish, fished in the Mariana Trench near Guam, at a depth of about $8{,}000\ m$. This small fish has a pink body, without

[4]https://scitechdaily.com/scientists-discover-dazzling-array-of-otherworldly-creatures-in-never-before-seen-underwater-canyons/

scales, and it would seem unlikely that it would be able to survive in such a hostile environment. In reality, it dominates this ecosystem, going deeper than any other fish and exploiting the absence of competitors, while devouring the abundant invertebrate prey that inhabit the pit. The fish's scientific name is *Pseudoliparis swirei*.

According to *NOAA*, the Mariana Trench is part of the "*Hades zone*" or "*Hadœl zone*" – the name was taken from "*Hades*", the Greek underworld. This region is composed mainly of marine trenches and extends at depths of approximately $6{,}000 - 11{,}000\ m$.

The snailfish appear well acclimated to these enormous pressures. There are more than *300* known species in the snailfish family. A group of researchers collected *38* specimens during expeditions carried out in 2014 and 2017, sampling at depths in the trench ranging from $6{,}900 - 8{,}000\ m$. The collected snailfish ranged in length between $89 - 235\ mm$, although their bodies shrank by about 10% during storage.

Pseudoliparis swirei, a type of snailfish that lives at the bottom of the Mariana Trench, are informally known as "*Hadœl*" snailfish because they live in the "*Hadœl zone*" of the ocean. Previous research has shown that there are many creatures living in the Trench, and snailfishes are the dominant predators. Furthermore, it is argued today that it is largely unknown how living organisms, particularly vertebrates, can survive and thrive in the darkness and high pressures of the "*Hadœl zone*". Some studies have used a remote-controlled underwater robot to explore the bottom of the pit and capture some specimens. They also studied the genetic sequencing of the genome to compare it to another type of snailfish that lives in tidal ponds.

The snailfish's skeleton is made of cartilage, which is much better suited to high pressures than bones. They discovered that the "*Hadœl*" snailfish had also lost part of its genome – those parts that give sensitivity to light and make sight possible. However, other parts of the genome related to sight were still intact, so the fish might still be able to see.

Unlike closely related shallow-sea species, *Pseudoliparis swirei* has transparent, unpigmented skin and scales, thin, incompletely ossified bones, a bloated stomach, and an unclosed skull. It therefore has phylogenetic characteristics that substantially differentiate it from the species that lives near the sea surface. The differentiation would have occurred approximately $20\ Ma$ ago.

But let us recall another very curious case. These are amphipods, small crustaceans found in most of the Earth's waters - and one species, *Hirondellea gigas*, lives in the Mariana Trench, at approximately $10{,}911\ m$ deep. At these extreme pressures, calcium carbonate ($CaCO_3$) – which makes up the shells of amphipods and many other marine animals – reacts with CO_2 to dissolve more easily in water. Thus, amphipods and many other marine animals remain limp, naked, and vulnerable. Owing to this reason, amphipods are generally not found deeper than approximately $5{,}000\ m$. The presence of *Hirondellea gigas* in the Mariana Trench was thus a mystery. Now, scientists from the *University of Tokyo* have discovered in 2019 that *Hirondellea gigas* survives thanks to the use of an aluminum (Al) shell.

Indeed, the exoskeleton of *Hirondellea gigas* contains Al on its surface. But Al is sparse in seawater. How did *Hirondella gigas* find this Al? The researchers found that the crustacean uses sugar-based chemicals in its gut to extract Al ions from the ocean floor mud, that it swallows while feeding on plant debris that rains down from the surface. They generate these sugar-based molecules from the sugars they harvest from these plant debris. In alkaline seawater, Al ions form a gel of aluminum hydroxide [$Al(OH)_3$]. This substance is commonly used to combat stomach acidity. That is, the $Al(OH)_3$ gel is a chemical protection that forms an impermeable barrier on the exoskeleton. *Hirondellea gigas* is therefore the first known organism that extracts Al from clay in order to live in the depths of the ocean.

We won't go into further details. Just notice the almost incredible ways that living forms can exploit to try to survive. All these discoveries show that, in addition to microorganisms, even large living forms can survive and evolve in an environment with no direct input from solar radiation. Furthermore, it would seem that these living forms derive from forms developed more on the surface and emigrated into the great depths.

Therefore, in principle, similar environments can also be found on other planetary bodies that enjoy a negligible amount of solar radiation. This implies that living forms can exist on other objects in the Solar System, although – unlike on Earth – they find no friendly "climate", and must remain permanently in the dark depths of some subsurface ocean.

Similar situations on other celestial bodies

First, we need to ascertain whether a celestial body - which is not a star - has endogenous heat that it exhales. As for the planets and their respective satellites, we have some remarkable information. At least two kinds of data are of specific interest.

Evidence is given by the surface morphology. If there is endogenous heat, this originates tectonic activity - i.e., similarly to what happens on Earth, large fractures form and some parts rise, etc. Some planetary bodies show this peculiar feature, while others appear as almost perfectly spherical spheres, almost like large billiard balls, at most with a few cracks due to surface temperature changes. That is, their surface resembles the surface of an enormous frozen swamp more than Earth's continents and mountains.

Other essential information is the presence of a magnetic field. If these bodies have a magnetic field, they should have a *TD*-dynamo, exactly like the Earth. To test this hypothesis we can calculate the "gradient" of the gravitational potential in which a given body is immersed (see above) and multiply this gradient by the diameter of the object. This product should be indicative of the difference in gravitational stress between different parts of the body. That is, this should be indicative of the intensity of the energy provided by the internal *TD*-dynamo. The objects where this product is largest should have a magnetic field. Others don't have it. The test really seems to work perfectly.

Furthermore, an internal *TD*-dynamo implies a substantial release of endogenous energy due to Joule heat (just as in an electric heater), therefore tectonic activity. Where there is a magnetic field, one must observe tectonic activity, and this test also works very well. We deduce that it is very probable, practically certain, that these objects have endogenous energy.

Other evidence, deduced from complicated models carried out by scientists at *NASA* and other institutions, insists on the probable existence of oceans submerged beneath the surface of many bodies in the Solar System. These bodies see the Sun just as we see a simple luminous star. But, in any case, the solar radiation is a negligible contribution to these bodies.

We must therefore reasonably assume that these planetary bodies should have life forms in their respective oceans beneath the surface. These life forms could be at least partly comparable to those in the depths of the Earth's ocean trenches.

We could also expect additional specifics. For instance, the Moon and Mars had an ancient *TD*-dynamo that later faded off and never came back on. Evidence of ancient volcanic activity on both the Moon and Mars - as well as flows of black fluids from the Martian soil resembling a flow of possible oil - are additional very interesting evidence.

Or Pluto is part of the Pluto/Charon binary system, which also has four tiny, rocky, dizzily rotating planets. Since these little planets are electrically charged (due to the electric charge transmitted by the solar wind), they are magnetized objects due to their rapid rotation. All of this seems to add up, as it explains morphological features that are otherwise unexplained. The Pluto/Charon binary system is therefore a wonderful, as well as incredible, natural laboratory to verify the expectations of the *TD*-dynamo!

But we cannot enter into these fascinating details here. Mother Nature never stops surprising and marveling us.

We can conclude that it is very plausible to hypothesize that life (and antilife) are endemic phenomena in matter (and antimatter, respectively) throughout the Universe.

Chapter 29

Conclusion

We began our journey like Jean-Jacques Rousseau's famous "*bon sauvage*", imagining how we could see the world without the *a priori* assumption of Newton's principles - which on the other hand have enormous merits for the development of engineering and applied science.

We introduced new hypotheses - very timidly - and almost ashamed of going blatantly against the flow. We thus built - with a considerable amount of disbelief - a "logical building" that progressively more and more surprised ourselves.

We understood how the perception of "time" is strictly dependent on the organization of the Universe according to the rules of the golden ratio, and how the golden ratio is intrinsically related to fractality. The Universe is in a perpetual search for an equilibrium that it never reaches. If it reaches equilibrium everything would end, and the entire Universe would be a still swamp where nothing happens. There would be no life, there would be no perception of "time", both by man and by any living form.

All this strictly implies irreversibility - in stark contrast to the Galilean teaching, which requires isolating individual natural subsets and looking for chains of cause and effect, on the assumption that every observed state is one of complete equilibrium. The Galilean criterion has immense merits for engineering and applied sciences, but it is based on an implicit assumption that is intrinsically opposed to irreversibility. This is a true logical bug.

We have discovered how the existence of antimatter seems can to be explained only by supposing that the truly absolute world is the timeless world β, which occasionally generates two symmetrical worlds, which are the complex conjugated image of each other. The world α is our world with matter, time, and entropy, and the world γ is the complex conjugated world with antimatter, antitime, and antientropy.

We also discovered how we must necessarily distinguish - in everything that we can objectively observe due to the empirical constraint - between the micro-cosmos and the "statistical" treatment of phenomena.

We have discovered how essential it is to acknowledge Mother Nature's organization into monads. Furthermore, it seemed clear to us to conclude that the laws of "statistics" - which hold in our macro-world of lotteries, election polls, etc. - really cannot hold within other monads. Thus, the much celebrated

Schrödinger equation is nothing more than the correct "statistical" treatment of the atom monad. Furthermore, as such, the same equation precisely cannot be applied to monads larger or smaller than the atom, such as, e.g., to the monad atomic nucleus.

We were finally able to enter the world of life in world α, and antilife in world γ, having however, to introduce a new concept of "statistics", taken from the sciences of the finance world, and we thus discovered the very important role of antifragility, thus how we can understand the fundamental role of autopoiesis.

To our ever greater surprise, we have realized that it appears more than reasonable to suppose that life is an endemic phenomenon in the Universe, and we have thus considered as experimental evidences - both in the depths of the Earth's oceans and the morphological characteristics of bodies in the Solar System - are consistent with this intriguing hypothesis. In fact, the life forms that exist in the various bodies scattered throughout the Universe correspond to the objective possibilities of survival and reproduction in order to allow autopoiesis. Indeed, in consideration of irreversibility, nothing can be eternal, but in order not to go extinct, everything must regenerate for being able to replace with new systems those which, due to the increase in entropy, can no longer exist and must die.

We thus imagined the existence of galaxies of matter and antigalaxies of antimatter – which is a hypothesis originally formulated by Hannes Olof Gösta Alfvén (1908–1995), Swedish physicist, Nobel Prize winner for Physics in 1970. Galaxies of matter and antigalaxies of antimatter sometimes annihilate upon entering the timeless world β - and in this way the entropy (which has grown disproportionately in galaxies) and the antientropy (which has similarly grown in antigalaxies) are reset.

An enormous phenomenon of materialization from the timeless world β will then be able to generate new galaxies and - symmetrically - new antigalaxies, complex conjugated to each other, thus completing a cycle that can be repeated indefinitely.

Furthermore, in the atemporal world β, it is not possible to define an extension of space, given that the concept of 3D space applies only in the world α or in the world γ, while it is not possible to define any space in the world β (see Chapter 23). Thus, the big-bang model, characterized by a huge explosion from a central point, makes no sense, as does the big-crunch model, characterized by a huge implosion at a single point.

Big-bang events and big-crunch events are measurable only in a 3D space in the α-world or the γ-world, but not in the timeless β-world. The galaxies and antigalaxies that we observe with our telescopes are generated at sites that, for observers of the α-world or the γ-world, are randomly located in 3D space. Cosmological models must therefore be fundamentally rethought.

Therefore, we should observe 50% of galaxies or antigalaxies that are moving away from us, and 50% of galaxies or antigalaxies that are approaching us. If the evidence of the so-called expansion of the Universe is correct, we must infer one of two possibilities. One possibility is that our galaxy is in a particular "bubble" that mainly "sees" receding galaxies and anti-galaxies. The other possibility is that when a complete miniphoton volley, i.e. one photon, travels very large distances, its wavelength increases (redshift). That is, it is as if the "wavelength" of the miniphotons within a complete volley that constitutes a photon "relaxes" when the photon comes from an extremely distant source. However, this last hypothesis is very *ad hoc*. The problem therefore remains open as does the hypothesis regarding the possibility of superluminal phenomena (see Chapter 22, Chapter 23, Chapter 24, and Chapter 25).

While this entire study was being edited in its orthographic details, the very prestigious American journal *Proceedings of the National Academy of Sciences* published a paper by Michael L. Wong et al. (of the *Earth and Planets Laboratory, Carnegie Institution for Science*, Washington, DC) where they formulate the "*Law of the increase of functional information*", a hypothesis that in some way - although starting from the laws of motion, gravity, electromagnetism, and thermodynamics - connects very vaguely to the unifying vision of the perspective of our journey.

In fact, we greatly broaden this perspective by explicitly including the entire Universe from the subnuclear scale up to galactic superclusters, and - no less importantly and above all - we get rid of the various paradoxes that have plagued theoretical physics for over a century. Furthermore, we also include phenomena linked to life in the Universe. The "*Law of the increase of functional information*" is thus generalized to the principle that can be called "*Law of antifragility in the universe*", where, through the golden ratio, the search for harmony is implied, linked to the tendency towards equilibrium, in accordance with the requirement of universal order.

We invite the reader to consider the nebulous scenario that we had before our eyes when, in Chapter 1, we defined the empirical constraint - and finally consider the surprising results of our journey, based only on considerations of common sense, of "simplicity", and of humility of in the face of Mother Nature. All of this seems almost incredible!

Thanks for reading through to the end.

"There are more things in heaven and earth, Horatio,
Than are dreamt of in your philosophy"

(*Hamlet*, Act I, Scene, IV, 166-167)

Acknowledgements

GPG fervently remembers with gratitude the late Prof. Helmut Moritz for his strong and sure encouragement when this study was taking its first steps, and also thanks the late Prof. Zbigniew Oziewicz for his support and encouragement along the way. The late Prof. Wilfried Schrőder was of great and friendly help when GPG published the 2005 monograph. Both authors GPG and BGG thank the co-authors of the multi-authored scientific investigation on which this volume is based, namely Martin T. Hovland, Bruce A. Leybourne, Stergios Pellis, Valentino Straser, Giorgio M. Gregori, and Anna Rita Simonelli whose contribution has been relevant while implementing several arguments. We also thank Willie Soon for substantial discussions, and Franco Ferdinand J. Bonavia for suggesting the title of the present volume.

Bibliography

Gregori, G.P., 2002. Galaxy – Sun – Earth relations. The origin of the magnetic field and of the endogenous energy of the Earth, with implications for volcanism, geodynamics and climate control, and related items of concern for stars, planets, satellites, and other planetary objects. A discussion in a prologue and two parts. *Beiträge zur Geschichte der Geophysik und Kosmischen Physik*, Band 3, Heft 3: pp. 1-471 [Available at http://ncgtjournal.com/additional-resources.html]

Gregori, G.P., 2005. Relativity, Quanta, Gravitation and Cosmology. A Discussion on the Cognitive Process in Theoretical Physics, *Sonderband Beitr. Gesch. Geophy. Kosm. Physik*, Science Edition. ISSN: 1615-2824, pp.: 219

Gregori, G.P., 2006f. Galaxy-Sun-Earth relations: the origin of the magnetic field and of the endogenous energy of the Earth, with implications for volcanism, geodynamics and climate control and related items of concern for stars, planets, satellites, and other planetary objects, *New Concepts Global Tectonics, Newsletters*, (38), 34-36

Gregori, G.P., 2007a. An axiomatic formulation of physics, IV: The time arrow, and overcoming action-at-a-distance and universal gravitation , *Tensor, N. S.*, 68, 301-325

Gregori, G.P., 2010. On the Pioneer anomaly and the Doppler effect. *Galilean Electrodynamics*, 21, (3): 43-52

Gregori, G.P., 2020. Climate change, security, sensors. *Acoustics*, 2: 474-504;DOI:10.3390/acoustics2030026.[https://www.mdpi.com/2624599X/2/3/26/htm]

Gregori, G.P., and B.A. Leybourne, 2025c. The solar cycle and MiniMax - Effects on other planets. In press in *New Concepts in Global Tectonics, Journal*

Gregori, G.P., and B.A. Leybourne, 2025e. The physics of electrical discharges – 1. Small-scale phenomena - Fog - atmospheric precipitation – BLs. In press in *New Concepts in Global Tectonics, Journal*

Gregori, G.P., and Wen-jie Dong, 1996. The correlation between the geomagnetic field reversals, the Hawaiian vulcanism and the motion of the Pacific plate. *Annali di Geofisica*, 39 (1): 49-65

Gregori, G.P., B.A. Leybourne, and J.R. Wright, 2025d. Generalized Cowling theorem and the Cowling dynamo. In press in *New Concepts in Global Tectonics, Journal*

Gregori, G.P., N. Accornero, B.G. Gregori, and M. Capozza, 2006. The foundations of physics, and the Bible. The thought and the brain. In W. Schröder (Ed.), *Natural science, philosophy and religion (Naturwissenschaft, Philosophie und Religion)*, Science Editions, Bremen-Roennebeck, pp:53-126

Jouis, C., and M. Orús-Lacort, 2021. Value of the golden ratio (number φ) knowing the side length of a square, preprint; DOI:10.13140/RG.2.2.20468.07045

Sagnac, G., 1913. On the proof of the reality of the luminiferous æther by the experiment with a rotating interferometer. *Comptes Rendus*, 157: 1410–1413

Sagnac, G., 1913a. The demonstration of the luminiferous æther by an interferometer in uniform rotation. *Comptes Rendus*, 157: 708–710

Tinè, S., 1987. Nuovi scavi nel santuario di Monte d'Accoddi, *Annali dell'Istituto Universitario Orientale di Napoli. Archeologia e Storia Antica*, IX: 9-22

Tinè, S., and A. Traverso, 1992a. Il restauro della struttura. In Tinè, S., and F. Lo Schiavo, (eds), 1992. *Monte d'Accoddi, 10 anni di scavi*. Istituto Italiano di Archeologia Sperimentale, and Soprintendenza Archeologica di Sassari e Nuoro, pp: xxxv-xl

Wong, M.L., C.E. Cleland, D. Arend, S. Bartlett, H.J. Cleaves, H. Demarest, A. Prabhu, J.I. Lunine, and R.M. Hazen, 2023. On the roles of function and selection in evolving systems, *Proceedings of the National Academy of Sciences*; DOI:10.1073/pnas.2310223120 europe books

www.ingramcontent.com/pod-product-compliance
Lightning Source LLC
LaVergne TN
LVHW070248220926
848482LV00043B/1021